NATURAL DISASTER RESEARCH, PREDICTION AND MITIGATION

# FEDERAL DISASTER ASSISTANCE AFTER THE 2005 AND 2008 GULF COAST HURRICANES

# NATURAL DISASTER RESEARCH, PREDICTION AND MITIGATION

Additional books in this series can be found on Nova's website under the Series tab.

Additional e-books in this series can be found on Nova's website under the e-book tab.

Natural Disaster Research, Prediction and Mitigation

# Federal Disaster Assistance After the 2005 and 2008 Gulf Coast Hurricanes

Madeline Payton
Editor

Copyright © 2014 by Nova Science Publishers, Inc.

**All rights reserved.** No part of this book may be reproduced, stored in a retrieval system or transmitted in any form or by any means: electronic, electrostatic, magnetic, tape, mechanical photocopying, recording or otherwise without the written permission of the Publisher.

For permission to use material from this book please contact us:
Telephone 631-231-7269; Fax 631-231-8175
Web Site: http://www.novapublishers.com

### NOTICE TO THE READER

The Publisher has taken reasonable care in the preparation of this book, but makes no expressed or implied warranty of any kind and assumes no responsibility for any errors or omissions. No liability is assumed for incidental or consequential damages in connection with or arising out of information contained in this book. The Publisher shall not be liable for any special, consequential, or exemplary damages resulting, in whole or in part, from the readers' use of, or reliance upon, this material. Any parts of this book based on government reports are so indicated and copyright is claimed for those parts to the extent applicable to compilations of such works.

Independent verification should be sought for any data, advice or recommendations contained in this book. In addition, no responsibility is assumed by the publisher for any injury and/or damage to persons or property arising from any methods, products, instructions, ideas or otherwise contained in this publication.

This publication is designed to provide accurate and authoritative information with regard to the subject matter covered herein. It is sold with the clear understanding that the Publisher is not engaged in rendering legal or any other professional services. If legal or any other expert assistance is required, the services of a competent person should be sought. FROM A DECLARATION OF PARTICIPANTS JOINTLY ADOPTED BY A COMMITTEE OF THE AMERICAN BAR ASSOCIATION AND A COMMITTEE OF PUBLISHERS.

Additional color graphics may be available in the e-book version of this book.

**Library of Congress Cataloging-in-Publication Data**

ISBN: 978-1-63117-886-3

*Published by Nova Science Publishers, Inc. † New York*

# CONTENTS

| | | |
|---|---|---|
| **Preface** | | vii |
| **Chapter 1** | Federal Disaster Assistance after Hurricanes Katrina, Rita, Wilma, Gustav, and Ike<br>*Bruce R. Lindsay and Jared Conrad Nagel* | 1 |
| **Chapter 2** | Disaster Relief Funding and Supplemental Appropriations for Disaster Relief<br>*Bruce R. Lindsay and Justin Murray* | 113 |
| **Chapter 3** | FEMA's Disaster Declaration Process: A Primer<br>*Francis X. McCarthy* | 145 |
| **Index** | | 177 |

# PREFACE

This book describes the various components of the Disaster Relief Fund (DRF), including what authorities have shaped it over the years; how FEMA determines the amount of the appropriation requested to Congress (pertaining to the DRF); and how emergency supplemental appropriations are requested. Information is also provided on funds appropriated in supplemental appropriations legislation to agencies other than the Department of Homeland Security (DHS).

Chapter 1 – This report provides information on federal financial assistance provided to the Gulf States after major disasters were declared in Alabama, Florida, Louisiana, Mississippi, and Texas in response to the widespread destruction that resulted from Hurricanes Katrina, Rita, and Wilma in 2005 and Hurricanes Gustav and Ike in 2008.

Congressional interest in Gulf Coast assistance has increased in recent years because of the significant amount of assistance provided to the region. Congress has also been interested in how the money has been spent, what resources have been provided to the region, and whether the money has reached the people and entities intended to receive the funds. The financial information is also useful for congressional oversight of the funds to identify the entities that have received the funds and to evaluate the overall effectiveness of the assistance. In addition, the information can help frame the congressional debate concerning federal assistance for current and future disasters.

The financial information for the 2005 and 2008 Gulf Coast storms is provided in two sections of this report:

1) Table 1 of Section I summarizes disaster assistance supplemental appropriations enacted into public law primarily for the needs associated with the five hurricanes, with the information categorized by federal department and agency; and
2) Section II contains information on the federal assistance provided to the five Gulf Coast states through the most significant federal programs, or categories of programs.

The financial findings in this report include:

- Congress has appropriated roughly $120.5 billion in hurricane relief for the 2005 and 2008 hurricanes in 10 supplemental appropriations statutes.
- The appropriated funds have been distributed among 11 departments, 3 independent agencies/entities, numerous sub-entities, and the federal judiciary.
- Congress appropriated almost half of the funds ($53 billion, or 44% of the total) to the Department of Homeland Security, most of which went to the Disaster Relief Fund (DRF) administered by the Federal Emergency Management Agency (FEMA).
- Congress targeted roughly 22% of the total appropriations (almost $27 billion) to the Department of Housing and Urban Development for community development and housing programs.
- Almost $25 billion was appropriated to Department of Defense entities: $15.6 billion for civil construction and engineering activities undertaken by the Army Corps of Engineers and $9.2 billion for military personnel, operations, and construction costs.
- FEMA has reported that roughly $5.9 billion has been obligated from the DRF after Hurricanes Katrina, Rita, and Wilma to save lives and property through mission assignments made to over 50 federal entities and the American Red Cross (see Table 19), $160.4 million after Hurricane Gustav through 32 federal entities (see Table 20), and $441 million after Hurricane Ike through 30 federal entities (see Table 21). In total, federal agencies obligated roughly $6.5 billion for mission assignments after the five hurricanes.
- The Small Business Administration approved almost 177,000 applications in the region for business, home, and economic injury

loans, with a total loan value of almost $12 billion (Table 31 and Table 32).
- The Department of Education obligated roughly $1.8 billion to the five states for elementary, secondary, and higher education assistance (Table 12).

This report also includes a brief summary of each hurricane and a discussion concerning federal to state cost-shares. Federal assistance to states is triggered when the President issues a major disaster declaration. In general, once declared the federal share for disaster recovery is 75% while the state pays for 25% of recovery costs. However, in some cases the federal share can be adjusted upward when a sufficient amount of damage has occurred, or when altered by Congress (or both). In addition, how much federal assistance is provided to states for major disasters is influenced not only by the declaration, but also by the percentage the federal government pays for the assistance. This report includes a cost-share discussion because some of these incidents received adjusted cost-shares in certain areas.

Since 2005 Congress has been interested in not only the amount of funding that has been directed to the Gulf Coast after the 2005 and 2008 hurricanes, but also in the wide range of activities and programs brought to bear to help the Gulf Coast states recover and prepare for future storms. This report summarizes the funds Congress directed to the area as well as the federal activities and programs that were put to use in response to the 2005 and 2008 hurricanes.

Chapter 2 – When a state is overwhelmed by an emergency or disaster, the governor may request assistance from the federal government. Federal assistance is contingent on whether the President issues an emergency or major disaster declaration. Once the declaration has been issued the Federal Emergency Management Agency (FEMA) provides disaster relief through the use of the Disaster Relief Fund (DRF), which is the source of funding for the Robert T. Stafford Emergency Relief and Disaster Assistance Act response and recovery programs. Congress appropriates money to the DRF to ensure that funding for disaster relief is available to help individuals and communities stricken by emergencies and major disasters (in addition, Congress appropriates disaster funds to other accounts administered by other federal agencies pursuant to federal statutes that authorize specific types of disaster relief).

Historically, the DRF is generally funded at a level that is sufficient for what are known as "normal" disasters. These are incidents for which DRF outlays are less than $500 million. When a large disaster occurs, funding for the DRF may be augmented through emergency supplemental appropriations. A supplemental appropriation generally provides additional budget authority during the current fiscal year to (1) finance activities not provided for in the regular appropriation; or (2) provide funds when the regular appropriation is deemed insufficient.

This methodology used to budget the DRF appears to have been altered since the passage of the Budget Control Act (P.L. 112-25). The Budget Control Act includes a series of provisions that directed the Office of Management and Budget (OMB) to annually calculate an "allowable adjustment" for disaster relief to the BCA's discretionary spending caps. That adjustment, if used, would make additional budget authority available for the federal costs incurred by major disasters declared under the Stafford Act beyond what is allowed in the regular discretionary budget allocation. The OMB calculation may have provided a mechanism that encourages a larger regular appropriation to the DRF. It is possible that larger DRF appropriations may reduce the need for supplemental appropriations.

Budgeting for disaster relief has been the subject of a great deal of debate. Some argue that more money should be appropriated in FEMA's DRF account in annual appropriations, while others maintain that augmenting the DRF through supplemental appropriations is preferable because it allows Congress to react directly to a particular situation. Others may argue that emergency supplemental appropriations are preferable for fiscal management reasons because an appropriation is not requested unless there is a real need for supplemental funding. Another argument is to revamp the budgetary process to fund disaster relief.

This report describes the various components of the DRF, including (1) what authorities have shaped it over the years; (2) how FEMA determines the amount of the appropriation requested to Congress (pertaining to the DRF); and (3) how emergency supplemental appropriations are requested. Information is also provided on funds appropriated in supplemental appropriations legislation to agencies other than the Department of Homeland Security (DHS). Aspects of debate concerning how disaster relief is budgeted are also highlighted and examined, and alternative budgetary options are summarized.

Chapter 3 – The Robert T. Stafford Disaster Relief and Emergency Assistance Act (referred to as the Stafford Act - 42 U.S.C. 5721 et seq.) authorizes the President to issue "major disaster" or "emergency" declarations before or after catastrophes occur. Emergency declarations trigger aid that protects property, public health, and safety and lessens or averts the threat of an incident becoming a catastrophic event. A major disaster declaration, issued after catastrophes occur, constitutes broader authority for federal agencies to provide supplemental assistance to help state and local governments, families and individuals, and certain nonprofit organizations recover from the incident.

The end result of a presidential disaster declaration is well known, if not entirely understood. Various forms of assistance are provided, including aid to families and individuals for uninsured needs and assistance to state and local governments and certain non-profits in rebuilding or replacing damaged infrastructure.

The amount of assistance provided through presidential disaster declarations has exceeded $140 billion. Often, in recent years, Congress has enacted supplemental appropriations legislation to cover unanticipated costs. While the amounts spent by the federal government on different programs may be reported, and the progress of the recovery can be observed, much less is known about the process that initiates all of this activity. Yet, it is a process that has resulted in an average of more than one disaster declaration a week over the last decade.

The disaster declaration procedure is foremost a process that preserves the discretion of the governor to request assistance and the President to decide to grant, or not to grant, supplemental help. The process employs some measurable criteria in two broad areas: Individual Assistance that aids families and individuals and Public Assistance that is mainly for repairs to infrastructure. The criteria, however, also considers many other factors, in each category of assistance, that help decision makers assess the impact of an event on communities and states.

Under current law, the decision to issue a declaration rests solely with the President. Congress has no formal role, but has taken actions to adjust the terms of the process. For example, P.L. 109-295 established an advocate to help small states with the declaration process. More recently, Congress introduced legislation, H.R. 3377, that would direct FEMA to update some of its criteria for considering Individual Assistance declarations.

Congress continues to examine the process and has received some recommendations for improvements. Given the importance of the decision, and the size of the overall spending involved, hearings have been held to review the declaration process so as to ensure fairness and equity in the process and its results.

In: Federal Disaster Assistance ...
Editor: Madeline Payton

ISBN: 978-1-63117-886-3
© 2014 Nova Science Publishers, Inc.

*Chapter 1*

# FEDERAL DISASTER ASSISTANCE AFTER HURRICANES KATRINA, RITA, WILMA, GUSTAV, AND IKE[*]

## Bruce R. Lindsay and Jared Conrad Nagel

### SUMMARY

This report provides information on federal financial assistance provided to the Gulf States after major disasters were declared in Alabama, Florida, Louisiana, Mississippi, and Texas in response to the widespread destruction that resulted from Hurricanes Katrina, Rita, and Wilma in 2005 and Hurricanes Gustav and Ike in 2008.

Congressional interest in Gulf Coast assistance has increased in recent years because of the significant amount of assistance provided to the region. Congress has also been interested in how the money has been spent, what resources have been provided to the region, and whether the money has reached the people and entities intended to receive the funds. The financial information is also useful for congressional oversight of the funds to identify the entities that have received the funds and to evaluate the overall effectiveness of the assistance. In addition, the information can help frame the congressional debate concerning federal assistance for current and future disasters.

---

[*] This is an edited, reformatted and augmented version of Congressional Research Service Publication, No. R43139, dated July 5, 2013.

The financial information for the 2005 and 2008 Gulf Coast storms is provided in two sections of this report:

3) **Table 1** of **Section I** summarizes disaster assistance supplemental appropriations enacted into public law primarily for the needs associated with the five hurricanes, with the information categorized by federal department and agency; and
4) **Section II** contains information on the federal assistance provided to the five Gulf Coast states through the most significant federal programs, or categories of programs.

The financial findings in this report include:

- Congress has appropriated roughly $120.5 billion in hurricane relief for the 2005 and 2008 hurricanes in 10 supplemental appropriations statutes.
- The appropriated funds have been distributed among 11 departments, 3 independent agencies/entities, numerous sub-entities, and the federal judiciary.
- Congress appropriated almost half of the funds ($53 billion, or 44% of the total) to the Department of Homeland Security, most of which went to the Disaster Relief Fund (DRF) administered by the Federal Emergency Management Agency (FEMA).
- Congress targeted roughly 22% of the total appropriations (almost $27 billion) to the Department of Housing and Urban Development for community development and housing programs.
- Almost $25 billion was appropriated to Department of Defense entities: $15.6 billion for civil construction and engineering activities undertaken by the Army Corps of Engineers and $9.2 billion for military personnel, operations, and construction costs.
- FEMA has reported that roughly $5.9 billion has been obligated from the DRF after Hurricanes Katrina, Rita, and Wilma to save lives and property through mission assignments made to over 50 federal entities and the American Red Cross (see **Table 19**), $160.4 million after Hurricane Gustav through 32 federal entities (see **Table 20**), and $441 million after Hurricane Ike through 30 federal entities (see **Table 21**). In total, federal agencies obligated roughly $6.5 billion for mission assignments after the five hurricanes.
- The Small Business Administration approved almost 177,000 applications in the region for business, home, and economic injury loans, with a total loan value of almost $12 billion (**Table 31** and **Table 32**).

- The Department of Education obligated roughly $1.8 billion to the five states for elementary, secondary, and higher education assistance (**Table 12**).

This report also includes a brief summary of each hurricane and a discussion concerning federal to state cost-shares. Federal assistance to states is triggered when the President issues a major disaster declaration. In general, once declared the federal share for disaster recovery is 75% while the state pays for 25% of recovery costs. However, in some cases the federal share can be adjusted upward when a sufficient amount of damage has occurred, or when altered by Congress (or both). In addition, how much federal assistance is provided to states for major disasters is influenced not only by the declaration, but also by the percentage the federal government pays for the assistance. This report includes a cost-share discussion because some of these incidents received adjusted cost-shares in certain areas.

Since 2005 Congress has been interested in not only the amount of funding that has been directed to the Gulf Coast after the 2005 and 2008 hurricanes, but also in the wide range of activities and programs brought to bear to help the Gulf Coast states recover and prepare for future storms. This report summarizes the funds Congress directed to the area as well as the federal activities and programs that were put to use in response to the 2005 and 2008 hurricanes.

## INTRODUCTION

This report provides a comprehensive summary of the federal financial assistance provided to the Gulf Coast states of Alabama, Florida, Louisiana, Mississippi, and Texas in response to the widespread destruction that resulted from Hurricanes Katrina, Rita, and Wilma in 2005 and Hurricanes Gustav and Ike in 2008.

The damages caused by the hurricanes are some of the worst in the history of the United States in terms of lives lost and property damaged and destroyed. The federal government played a significant role in the response to the hurricanes and Congress appropriated funds for a wide range of activities and efforts to help the Gulf Coast states recover and rebuild from the storms. In addition, Congress appropriated a significant amount of funds to reduce or eliminate the impacts of future storms.

Congressional interest in Gulf Coast assistance has increased in recent years because of the significant amount of assistance provided to the region. Congress has also been interested in how the money has been spent, what resources have been provided to the region, and whether the money has reached the people and entities intended to receive the funds. The financial information is also useful for congressional oversight of the funds to identify the entities that have received the funds and to evaluate the overall effectiveness of the assistance. In addition, the information can help frame the congressional debate concerning federal assistance for current and future disasters.

The financial information provided in this report includes a summary of appropriations provided to the Gulf Coast states by Congress in response to the 2005 and 2008 hurricanes. In addition, when available, hurricane specific and state specific funding information is provided by federal entity.

# BACKGROUND[1]

The 2005 hurricane season was a record breaking season for hurricanes and storms. There were 13 hurricanes in 2005, breaking the old record of 12 hurricanes set in 1969.[2] The 2005 season also set a record for the number of category 5 storms (three) in a season.[3] Most of the damaging effects caused by the hurricanes were experienced in the Gulf Coast states of Louisiana, Arkansas, Florida, Mississippi, and Texas. The 2008 hurricane season was also an active hurricane season that caused additional damage in the Gulf Coast.

## Hurricane Katrina

On August 23, 2005, Hurricane Katrina began about 200 miles southeast of Nassau in the Bahamas as a tropical depression. It became a tropical storm the following day. On August 24-25, 2005, the storm moved through the northwestern Bahamas and then turned westward toward southern Florida. Katrina became a hurricane just before making landfall near the Miami-Dade/Broward county line during the evening of August 25, 2005. The hurricane moved southwestward across southern Florida into the eastern Gulf of Mexico on August 26, 2005.

Katrina then strengthened significantly, reaching Category 5 intensity on August 28. On August 29, 2005, Hurricane Katrina made landfall in southern Plaquemines Parish, Louisiana. The storm affected a broad geographic area—stretching from Alabama, across coastal Mississippi, to southeast Louisiana. Hurricane Katrina was reported as a category 4 storm when it initially made landfall in Louisiana, but was later downgraded to a category 3 storm. Even as a category 3 storm, Hurricane Katrina was one of the strongest storms to impact the U.S. Gulf Coast. The force of the storm was significant. The winds to the east of the storm's center were estimated to be nearly 125 mph.[4]

The Gulf Coast has had a history of devastating hurricanes, but Hurricane Katrina was singular in many respects. Approximately 1.2 million people evacuated from the New Orleans metropolitan area.[5] While the evacuation helped to save lives, over 1,800 people died in the storm.[6] In addition, Hurricane Katrina destroyed or made uninhabitable an estimated 300,000 homes[7] and displaced over 400,000 citizens.[8] Economic losses from the storm were estimated to be between $125 billion and $150 billion.[9]

## Hurricanes Rita and Wilma

Two other hurricanes made landfall in the Gulf Coast shortly after Hurricane Katrina that added to recovery costs and impeded recovery efforts. On September 24, 2005, Hurricane Rita made landfall on the Texas and Louisiana border as a category 3 storm. Rita also hit parts of Arkansas and Florida. Hurricane Rita caused widespread property damage to the Gulf Coast; however, there were few deaths or injuries reported.[10] Rita produced rainfalls of 5 to 9 inches over large portions of Louisiana, Mississippi, and eastern Texas, with isolated amounts of 10 to 15 inches.[11] In addition, storm surge flooding and wind damage occurred in southwestern Louisiana and southeastern Texas, with some surge damage occurring in the Florida Keys.[12]

On October 24, 2005, Hurricane Wilma made landfall as a Category 3 hurricane in Cape Romano, Florida. The eye of Hurricane Wilma crossed the Florida Peninsula and then moved into the Atlantic Ocean north of Palm Beach.[13] Hurricane Wilma killed five people in Florida and caused widespread property damage in the Gulf Coast region.

## Hurricanes Gustav and Ike

In 2008, the Gulf Coast was once again affected by storms that caused billions of dollars in additional damage. On September 1, 2008, Hurricane Gustav made landfall near Cocodrie, Louisiana, as a category 2 storm, then swept across the region causing damages in Alabama, Florida, Mississippi, and Texas. Gustav produced rains over Louisiana and Arkansas that caused moderate flooding along many rivers, and is known to have produced 41 tornadoes: 21 in Mississippi, 11 in Louisiana, 6 in Florida, 2 in Arkansas, and 1 in Alabama.[14]

Hurricane Ike made landfall as a category 2 storm near Galveston, Texas, on September 13, 2008, with maximum sustained winds of 110 mph. The hurricane weakened as it moved inland across eastern Texas and Arkansas. Hurricane Ike's storm surge devastated the Bolivar Peninsula of Texas, and surge, winds, and flooding from heavy rains caused widespread damage in other portions of southeastern Texas, western Louisiana, and Arkansas and killed 20 people in these areas.[15] Additionally, as an extratropical system over the Ohio Valley, Ike was directly or indirectly responsible for 28 deaths and more than $1 billion in property damage in areas outside of the Gulf Coast.[16]

## Historical Perspective[17]

The deaths and damages caused by the 2005 and 2008 hurricanes rank them among the worst disasters in U.S. history. **Figure 1** provides data from past, large-scale incidents that have occurred in the United States to provide context to the devastation wrought by the 2005 and 2008 hurricanes. Damages for Hurricanes Katrina, Rita, and Wilma (KRW) are combined together in **Figure 1**.

While these comparisons help to illustrate the scale of devastation from one disaster to another, it is important to note that all disasters, and especially disasters of the magnitude of Hurricanes Katrina, Rita, Wilma, Gustav, and Ike, are produced by a set of unique circumstances that result in an equally unique set of needs that may lead to assistance from the federal government.

Two major concepts are typically considered when comparing the need for federal assistance following disasters. First, because of the federalism principles of emergency management—that the federal government generally provides assistance to supplement the work of state, tribal, and local governments only after they become overwhelmed and only at their request— the varying capabilities of a state/tribal/local government can change the types and scope of assistance provided by the federal government. This issue was discussed by the Administrator of FEMA in recent congressional testimony on Hurricane Sandy. In reference to the denial of an application for one form of disaster assistance (individual assistance), Administrator Fugate explained that decisions to provide federal assistance are based not upon the need of any particular individual, but upon the need of the state as a whole and whether the state is capable of addressing that need without federal assistance.[18]

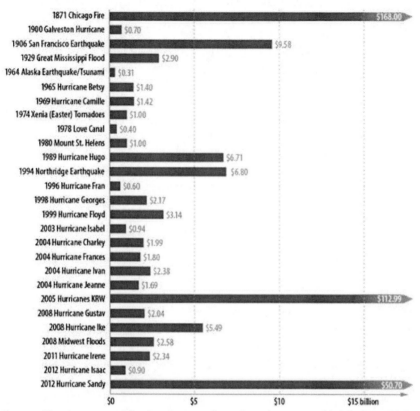

Source: The data sources for the above analyses have been assembled from multiple governmental sources to provide a rough comparison of disasters. These sources can be located in Appendix A. The data on damages from these sources are subject to variation and should not be viewed as definitive.

Figure 1. Examples of Most Expensive U.S. Disasters, 1871 to 2012. (2012 Dollars; in Billions).

Second, the relative levels of federal assistance required for each disaster depend on the proportional impact to various sectors of the community. For example, a particular disaster may destroy one community's business district and overwhelm the ability of the state to respond to that impact, while another may significantly damage the majority of the community's public facilities. In the first disaster, the assistance from the federal government may be noteworthy for the relatively large amount of loan assistance provided by the Small Business Administration, while the second disaster may be noteworthy for the relatively large amount of assistance provided through the FEMA's Public Assistance (PA) program.

Some additional disaster-specific factors that may inhibit the usefulness of making generalized disaster-to-disaster comparisons include:

- the density and socioeconomic status of the affected population;
- the percentage of properties and private/public losses that were insured, and the adequacy of the insurance coverage; and
- the number of jurisdictions affected by the disaster, and whether these jurisdictions span multiple states requiring greater federal coordination of the response and recovery effort.

## INFORMATION CATEGORIES AND DATA COLLECTION METHODS

The following two sections provide funding data and narratives describing the assistance that was provided to the Gulf Coast in response to the 2005 and 2008 hurricane seasons. Section I presents funding provided to the five Gulf Coast states (Alabama, Florida, Louisiana, Mississippi, and Texas) after Hurricanes Katrina, Rita, Wilma, Gustav, and Ike. Funding amounts were compiled by CRS analysts who reviewed legislative texts of supplemental appropriations. The amounts are disaggregated by federal entity and sub-entity, insofar as possible and applicable. The data are based on the analysts' interpretations of disaster assistance. Some data were excluded from Section I because CRS analysts found that the data either were too ambiguous or covered disasters not limited to the Gulf Coast. Certain amounts pertaining to a range of disasters were included, however, because CRS analysts determined that most of the funds went to the Gulf Coast states.

Section II presents funding by federal agency. The amounts reported may reflect expenditures, obligations, allocations, or appropriations. The data in this section are not based solely on those in Section I. Rather, the data in Section II were derived from a variety of authoritative sources, including agency websites, CRS experts who received information directly from agencies, and governmental reports. Section II presents funding information by federal entity and includes a narrative summarizing each agency's disaster assistance efforts. The sections also provide the authorities that authorized the activities that were provided. When possible, funding data are provided in tabular form.

It should be noted that the data on appropriations in Section I, Table 1, are not directly comparable to funding data in Section II. The former were drawn solely from the public laws cited in the source note to Table 1. The data in Section II were obtained, as cited in each subsection, from a range of published and unpublished sources, and include various fiscal years.

## CAVEATS AND LIMITATIONS

Funding data on federal (and non-federal) assistance are not systematically collected. Given the absence of comprehensive federal information on disaster assistance, the data provided in this report should only be considered as an approximation, and should not be viewed as definitive.

In addition to the above, the following caveats apply to this report:

- It is difficult to identify all of the federal entities that provide disaster relief because many federal entities provide aid through a wide range of programs, not necessarily through those designated specifically as "disaster assistance" programs.[19]
- Because data on federal (and non-federal) assistance are not systematically collected, funding data were drawn from a wide-range of sources including published and unpublished data that have been collected at different times and under inconsistent reporting methods.
- Following the exodus of thousands of residents from the Gulf Coast states after Hurricane Katrina in 2005, many other states received federal assistance to cope with the influx of those seeking aid. The aid provided to the states outside the Gulf Coast is not discussed in this report.
- The appropriations language reviewed for **Section I** usually designates funds to a federal entity for a range of disasters without identifying how much funding is to be disbursed to each incident. For example, P.L. 110-329, signed into law on September 30, 2008, provided funds for several disasters that occurred in 2008, including Hurricanes Gustav and Ike, wildfires in California, and the Midwest floods. Determining the funding amounts directed toward each

individual disaster is difficult, if not impossible, unless the legislative text specifies these amounts. An additional difficulty occurs in tracking funding at the agency level because appropriations might be made, not to specific entities, but to budget accounts, and then allocated for specified purposes.
- The degree of transparency in reporting funding levels for disaster relief varies tremendously among federal entities. As an example, Congress requires the Federal Emergency Management Agency to submit monthly status reports on the Disaster Relief Fund (DRF).[20] The DRF is FEMA's disaster assistance account. The DRF is used to fund existing recovery projects (including reimbursements to other federal agencies for their work) and provide funding for future emergencies and disasters as needed. The DRF reports must detail obligations, allocations, and expenditures for Hurricanes Katrina, Rita, and Wilma. This requirement has not been extended to other agencies, and scant data exist, particularly on a state-bystate basis, on other federal funding for emergencies and major disasters.
- Appropriations may be subject to transfers or rescissions after enactment of appropriations statutes. It is possible that such emendations to the initial appropriations have not been identified in this research. Accordingly, the summary totals for appropriations should be interpreted as illustrative as opposed to definitive, and used with caution.

## SECTION I: SUMMARY OF GULF COAST DISASTER SUPPLEMENTAL APPROPRIATIONS

Table 1 presents data on the appropriations enacted after Hurricanes Katrina, Rita, Wilma, Gustav, and Ike from FY2005 to FY2009, by federal entity and sub-entity, when possible and applicable. As mentioned earlier, in many cases funding for disaster relief is appropriated for multiple incidents. Therefore, Table 1 may include data on appropriations that also provided funding for non-Gulf Coast incidents. Some appropriations designated for a range of disasters were excluded, however, in an attempt to avoid artificially inflating the amount of funding directed to the Gulf Coast for hurricane relief.

Since FY2005, at least 10 appropriations bills have been enacted to address widespread destruction caused by the 2005 and 2008 Gulf Coast hurricanes. These appropriations consisted of eight emergency supplemental appropriations acts, one reconciliation act, and one continuing appropriations resolution.[21] In addition to these statutes that specifically identify the hurricanes or the Gulf Coast states, it is likely that regular appropriations legislation also provided assistance to the Gulf Coast. Because these statutes did not specify that they were providing such assistance, regular appropriations are not included in Table 1.

**Table 1. Estimated Gulf Coast Supplemental Appropriations for Hurricanes Katrina, Rita, Wilma, Gustav, and Ike (Disaster-Related Supplemental Appropriations by Department/Agency; Nominal Dollars in Millions)**

| Department/Agency/Program | Estimated Appropriation |
|---|---|
| **DEPARTMENT OF AGRICULTURE** | |
| Emergency Forestry Conservation Reserve Program | $82 |
| Agricultural Research Service | $39 |
| Emergency Conservation Program | $73 |
| Farm Service Agency | $242 |
| Executive Operations | $60 |
| Food and Nutrition Service Commodity Assistance | $10 |
| Forest Service | $77 |
| Inspector General | * |
| Natural Resources Conservation Service | $351 |
| Other Emergency Appropriations | * |
| Rural Housing Service | $90 |
| Rural Utility Service | $53 |
| Subtotal | $1,077 |
| **DEPARTMENT OF COMMERCE** | |
| Department of Commerce (non specified) | $400 |
| National Oceanic and Atmospheric Administration | $85 |
| Marine Fishery Emergency Assistance Program | $260 |
| Subtotal | $745 |
| **DEPARTMENT OF DEFENSE (MILITARY)** | |
| Military Personnel | $540 |
| Operations and Maintenance | $3,684 |
| Procurement | $2,850 |
| Research, Development, Test, and Evaluation | $54 |
| Military Construction and Family Housing | $1,785 |
| Management Funds | $66 |
| Other Defense | $236 |

## Table 1. (Continued)

| Department/Agency/Program | Estimated Appropriation |
|---|---|
| Subtotal | $9,215a |
| **DEPARTMENT OF DEFENSE (CIVIL)** | |
| Army Corps of Engineers Construction | $4,951 |
| Flood Control and Coastal Emergencies | $9,926 |
| Flood Damage Construction for FEMA | * |
| Mississippi River and Tributaries | $154 |
| General Expenses | $3 |
| Investigations | $43 |
| Operations and Maintenance | $516 |
| Subtotal | $15,593 |
| **DEPARTMENT OF EDUCATION** | |
| Office of Elementary and Secondary Education | $1,689 |
| Office of Postsecondary Education | $292 |
| Subtotal | $1,981 |
| **DEPARTMENT OF HEALTH AND HUMAN SERVICES** | |
| Health Resources and Services Administration | $4 |
| Administration for Children and Families | $1,240 |
| Centers for Disease Control and Prevention | $8 |
| Centers for Medicare and Medicaid Services | $2,000 |
| Subtotal | $3,252 |
| **DEPARTMENT OF HOMELAND SECURITY** | |
| Department of Homeland Security (non-specified) | $9,157 |
| Customs and Border Protection | $52 |
| Federal Emergency Management Agency | $44,083b |
| Immigration and Customs Enforcement | $13 |
| Office of Domestic Preparedness | $10 |
| Office of Inspector General | $2 |
| United States Coast Guard | $487 |
| United States Secret Service | $4 |
| Subtotal | $52,711 |
| **DEPARTMENT OF HOUSING AND URBAN DEVELOPMENT** | |
| Community Development Block Grants | $26,200 |
| Rental Assistance/Section 8 Vouchers | $555 |
| Supportive Housing | $73 |
| Public Housing Repair | $15 |
| Office of Inspector General | $7 |
| Subtotal | $26,850 |
| **DEPARTMENT OF THE INTERIOR** | |
| Department of the Interior | $210 |

| Department/Agency/Program | Estimated Appropriation |
|---|---|
| Bureau of Reclamation | $9 |
| Mineral Management Service | $31 |
| National Park Service | $117 |
| National Park Service Historical Preservation Fund | * |
| U.S. Fish and Wildlife Service | $162 |
| U.S. Geological Survey | $16 |
| Subtotal | $545 |
| **DEPARTMENT OF JUSTICE** | |
| Bureau of Alcohol, Tobacco, Firearms and Explosives | $20 |
| Drug Enforcement Administration | $10 |
| Federal Bureau of Investigation | $45 |
| Federal Prison System | $11 |
| Legal Activities | $18 |
| Office of Justice Programs | $175 |
| U.S. Marshals Service | $9 |
| Subtotal | $288 |
| **DEPARTMENT OF LABOR** | |
| Job Corps | $16 |
| Employment and Training Administration | $125 |
| Subtotal | 141 |
| **DEPARTMENT OF TRANSPORTATION** | |
| Department of Transportation (non-specified) | $722 |
| Federal Aviation Administration | $41 |
| Federal Highway Administration | $2,751 |
| Federal Transportation Administration Grants | * |
| Maritime Administration | $8 |
| Subtotal | $3,522 |
| **DEPARTMENT OF VETERANS AFFAIRS** | |
| Department Administration | $62 |
| Veterans Health Administration | $198 |
| Major Construction- Medical Facilities | $918 |
| Subtotal | $1,178 |
| **ARMED FORCES RETIREMENT HOME** | |
| Armed Forces Retirement Home | $242 |
| Subtotal | $242 |
| **ENVIRONMENTAL PROTECTION AGENCY** | |
| Environmental Protection Agency (non-specified) | $21 |
| Subtotal | $21 |
| **GENERAL SERVICES ADMINISTRATION** | |
| General Services Administration (non-specified) | $75 |
| Subtotal | $75 |

## Table 1. (Continued)

| Department/Agency/Program | Estimated Appropriation |
|---|---|
| **SMALL BUSINESS ADMINISTRATION** | |
| Small Business Administration (non-specified) | $2,279 |
| Disaster Loans Program Account | $441 |
| Inspector General | $5 |
| Subtotal | $2,725 |
| **THE JUDICIARY** | |
| The Federal Judiciary (non specified) | $18 |
| Subtotal | $18 |
| **NATIONAL AERONAUTICS AND SPACE ADMINISTRATION** | |
| National Aeronautics and Space Administration (non-specified) | $385 |
| Exploration Capabilities as a Consequence of Katrina | * |
| Subtotal | $385 |
| **CORPORATION FOR NATIONAL AND COMMUNITY SERVICE** | |
| Corporation for National and Community Service | $10 |
| Subtotal | $10 |
| Grand Total | $121,701 |

Source: Data derived from CRS database of appropriations. Statutes include: P.L. 109-61, P.L. 109-62, P.L. 109-148, P.L. 109-171, P.L. 109-234, P.L. 110-28, P.L. 110-116, P.L. 110-252, P.L. 110-329, and P.L. 111-32. This table does not take into consideration any rescissions applied after Congress appropriated these funds.

Notes: * Signifies appropriation of less than $1 million. Cells marked as "non-specified" indicate appropriations funded to a department generally.

[a] This figure represents the amount appropriated after rescission of funds; it does not reflect that $1.5 billion of these funds expired in FY2006 or were transferred for other purposes.

[b] P.L. 109-62 (119 Stat. 1991) appropriated $50 billion for the Disaster Relief Fund. P.L. 109-148 (119 Stat. 2790) rescinded $23.4 billion of those funds.

# SECTION II. AGENCY-SPECIFIC INFORMATION ON GULF COAST HURRICANE FEDERAL ASSISTANCE

In the course of this research, CRS identified 11 federal departments, 4 federal agencies (or other entities), and numerous sub-entities, programs, and activities that supplied over $120.5 billion in federal assistance to the Gulf Coast states after the major hurricanes of 2005 (Katrina, Rita, and Wilma) and

2008 (Gustav and Ike). **Section II** provides information on the most significant programs, or categories of programs, through which the aid was provided. Each narrative contains a summary of activities of each federal entity providing disaster relief. When possible, the information is presented in tabular form and is disaster and state specific. Unless otherwise specified, all figures are stated in nominal dollars.

As mentioned earlier, the data in **Section II** may not correspond to the emergency funds appropriated by Congress for hurricane relief purposes specified in **Section I**. Reasons for the difference include the following:

- the tables in **Section II** present information from a variety of funding measures, including obligations, allocations, and expenditures;[22]
- some funds made available may have been reallocated or deobligated from other purposes; and
- money from accounts that did not terminate at the end of a fiscal year (known as no-year accounts) may have been allocated to the Gulf Coast states.

## Department of Agriculture[23]

The U.S. Department of Agriculture (USDA) provides a variety of disaster assistance for hurricanes and other natural disasters. For the hurricanes covered in this report, the bulk of the department's funding has been disaster payments to producers who suffered production losses and funding for land rehabilitation programs for cleanup and restoration projects, primarily under P.L. 109-234 and through other authorities. The total USDA outlays were $792 million for disaster relief following Hurricanes Katrina, Rita, and Wilma (**Table 2**). For these three hurricanes, USDA also paid an additional $110 million in farm disaster benefits to farmers in the Gulf States under various Farm Service Agency indemnity and grant programs, using funds allocated from USDA's "Section 32" Program (see "Farm Service Agency" section below).[24]

Hurricane-related support by individual agency for the 2005 and 2008 hurricanes is described in separate sections below. State-specific data are provided where available.

## Table 2. Disaster Relief Funding by the U.S. Department of Agriculture for 2005 Gulf Coast Hurricanes
### (Dollars in Thousands)

| Department of Agriculture | Budget Authority | Obligations | Outlays |
|---|---|---|---|
| Agricultural Research Service | $39,000 | $38,000 | $37,000 |
| Farm Service Agency | | | |
| Disaster payments-crop/livestock losses (excludes Section 32) | $132,300 | $132,300 | $132,300 |
| Emergency Forestry Conservation Reserve Program (EFCRP) | $81,800 | $81,800 | $68,600 |
| Emergency Conservation Program (ECP) | $84,700 | $73,400 | $44,800 |
| Food and Nutrition Service | $10,000 | $10,000 | $9,000 |
| Forest Service | $77,000 | $77,000 | $77,000 |
| Office of Inspector General | $445 | $445 | $445 |
| Natural Resources Conservation Service (NRCS) | $351,000 | $300,000 | $287,000 |
| Rural Housing Service | $128,000 | $101,000 | $63,000 |
| Rural Utilities Service | $53,000 | $34,000 | $14,000 |
| Working Capital Fund | $60,000 | $59,000 | $59,000 |
| Total | $1,017,245 | $906,945 | $792,145 |

Source: Budget Data Request No. 11-31 requested June 27, 2011, and provided July 20, 2011. Submission by Office of Budget and Program Analysis, U.S. Department of Agriculture, to Office of Management and Budget.

Notes: Figures are for Hurricanes Katrina, Rita, and Wilma in support of Gulf Coast recovery efforts and include disaster payments made under P.L. 109-234. Excludes disaster payments made under Section 32 (see Table 3) and disaster payments made under the 2008 farm bill (see Table 4).

### *Agricultural Research Service*

The Agricultural Research Service (ARS) is USDA's chief scientific research agency. Under P.L. 109-234, USDA received funding for cleanup and salvage efforts at the ARS facility in Poplarville, Mississippi, and the Southern Regional Research Center in New Orleans, Louisiana. Total outlays were $37 million for the 2005 hurricanes provided under P.L. 109-234 and through reallocations from existing funds.

### *Farm Service Agency*

The mission of the Farm Service Agency (FSA) is to serve farmers, ranchers, and agricultural partners through the delivery of agricultural support programs. Besides administering general farm commodity programs, FSA administers disaster payments for crop and livestock farmers who suffer losses from natural disasters. Following the 2005 hurricanes, producer benefits were provided under five new programs created by USDA for tropical fruit, citrus, sugarcane, nursery crops, fruits and vegetables, livestock death, feed losses, and dairy production and spoilage losses. These included the Hurricane Indemnity Program (HIP), Livestock Indemnity Program (LIP), Feed Indemnity Program (FIP), Dairy Disaster Assistance Payment Program (DDAP), and an Aquaculture Grant Program (AGP). Payments under the Tree Indemnity Program (TIP) were provided to eligible owners of commercially grown fruit trees, nut trees, bushes, and vines producing annual crops that were lost or damaged. Total outlays for 2005 hurricanes to the Gulf States were $132 million under P.L. 109-234 and $110 million under "Section 32" (see **Table 3** for Section 32 data). Section 32 is a permanent appropriation (originating from P.L. 74-320) that supports a variety of USDA activities, including disaster relief, federal child nutrition programs, and surplus commodity purchases.

**Table 3. 2005 Hurricane Disaster Relief Payments for Crops and Livestock by State**
**(Dollars in Thousands)**

|  | Alabama | Florida | Louisiana | Mississippi | Texas | Total |
|---|---|---|---|---|---|---|
| Hurricane Indemnity Program (HIP) | $3,001 | $30,488 | $3,012 | $2,059 | $259 | $38,819 |
| Tree Indemnity Program (TIP) | $602 | $17,574 | $374 | $776 | $28 | $19,354 |
| Feed Indemnity Program (FIP) | $898 | $1,715 | $1,041 | $1,156 | $27 | $4,837 |
| Livestock Indemnity Program (LIP) | $265 | $601 | $19,056 | $2,121 | $242 | $22,285 |
| Aquaculture Grant Program (AGP) | $5,038 | $3,663 | $4,513 | $10,763 | $713 | $24,690 |
| **Total** | **$9,804** | **$54,041** | **$27,996** | **$16,875** | **$1,269** | **$109,985** |

Source: USDA/Farm Service Agency, November 6, 2012.
Notes: Latest payment data available from USDA for 2005 hurricanes, as of February 23, 2007; the above programs were administered by FSA with funding allocated from USDA's "Section 32" Program.

Following Hurricanes Gustav and Ike in 2008, payments were provided to qualifying producers under five nationwide agricultural disaster programs authorized in the Food, Conservation, and Energy Act of 2008 (P.L. 110-246, 2008 Farm Bill). Under the largest disaster program, Supplemental Revenue Assistance Payments Program (SURE), the combined payments for Alabama, Florida, Louisiana, Mississippi, and Texas totaled $285 million in 2008 for a variety of natural disaster losses, not just hurricane damage (**Table 4**). Payments for these states under the other four programs (three livestock-related programs and the Tree Assistance Program (TAP)) totaled $66 million.

**Table 4. 2008 Agricultural Disaster Relief Program Payments by State (Dollars in Thousands)**

|  | Alabama | Florida | Louisiana | Mississippi | Texas | Total |
|---|---|---|---|---|---|---|
| Supplemental Revenue Assistance Payments Program (SURE) | $5,005 | $12,932 | $13,068 | $4,993 | $248,993 | $284,991 |
| Livestock Forage Program (LFP) | $9,002 | $2,688 | - | - | $40,182 | $51,872 |
| Livestock Indemnity Program (LIP) | $34 | $64 | $1,301 | $91 | $6,359 | $7,849 |
| Emergency Assistance for Livestock, Honeybees and Farm-Raised Fish Program (ELAP) | $81 | $2,918 | $776 | $10 | $659 | $4,444 |
| Tree Assistance Program (TAP) | - | $1,798 | < $1 | - | $141 | $1,939 |
| **Total** | **$14,122** | **$20,400** | **$15,145** | **$5,094** | **$296,334** | **$351,095** |

Source: Farm Service Agency, U.S. Department of Agriculture, November 6, 2012.
Notes: Programs were authorized under the Food, Conservation, and Energy Act of 2008 (P.L. 110-246, "2008 farm bill"). Payments as of November 6, 2012, and made for a variety of natural disaster losses that included more than just hurricane damage.

FSA also administers two land rehabilitation disaster programs: (1) the Emergency Forestry Conservation Reserve Program (EFCRP),[25] which compensates private, nonindustrial forest landowners who experienced losses from hurricanes in calendar year 2005, for temporarily retiring their land; and (2) the Emergency Conservation Program (ECP),[26] which provides emergency funding and technical assistance for farmers and ranchers to rehabilitate farmland damaged by natural disasters.

For the 2005 hurricanes, Congress provided $82 million in budget authority for EFCRP and $84.7 million in budget authority for ECP. Of the $84.7 million in budget authority for ECP, FSA obligated $70 million. Previously unobligated funds from 2005 hurricane recovery efforts were reprogrammed in 2009 under P.L. 111-32 to be used for current disasters, including hurricanes. On July 14, 2009, USDA announced $71 million in ECP funding, which included the 2005 reprogrammed funds, for repairing farmland damaged by natural disasters, including the hurricanes that occurred in 2008. Of the five hurricane states, Texas received the largest allocation ($11 million) to address 2008 hurricane restoration efforts.

## *Food and Nutrition Service*

The Food and Nutrition Service (FNS) administers several programs that are crucial in hurricane relief efforts.[27] These include the Supplemental Nutrition Assistance Program or "SNAP" (the 2008 farm bill, P.L. 110-246, originally called the Food Stamp Program), child nutrition programs, and federally donated food commodities delivered through relief organizations. Existing laws authorize USDA to change eligibility and benefit rules to facilitate emergency aid. Food assistance benefits provided under FNS programs funded at least $1 billion worth of support directly due to Hurricanes Katrina, Rita, and Wilma. These required no additional appropriations because the benefits are entitlements.

Other than a small one-time increase in appropriations, in P.L. 109-148, to replenish some commodity stocks used for hurricane-relief purposes, no significant action was taken for hurricane relief or to pay for commodity distribution costs. This is because funding and federally provided food commodities were generally available without a need for a large appropriation.

## *Natural Resources Conservation Service*

The Natural Resources Conservation Service (NRCS) helps America's private land owners and managers conserve soil, water, and other natural resources. Following natural disasters, NRCS works with FEMA, state and federal agencies, and local units of government to conduct post-disaster cleanup and restoration projects. NRCS administers the Emergency Watershed Protection (EWP) Program,[28] which assists landowners and operators in implementing emergency recovery measures for slowing runoff and preventing erosion to relieve imminent hazards to life and property created by a natural disaster that causes a sudden impairment of a watershed. In the wake of hurricane events, NRCS staff also assesses the demand and requirement for the disposal of animal carcasses, through authority delegated by FEMA. As of November 29, 2012, NRCS had obligated approximately $299 million. State EWP data for the 2005 and 2008 hurricanes are provided in **Table 5** below.

**Table 5. Disaster Relief Funding through the Emergency Watershed Protection Program**
**(Dollars in Thousands)**

| Hurricane | Alabama | Florida | Louisiana | Mississippi | Tennessee | Texas | Total |
|---|---|---|---|---|---|---|---|
|  |  |  | Obligations |  |  |  |  |
| Katrina | $21,300 | $7,200 | $44,900 | $114,200 | $400 | - | $188,000 |
| Rita | - | - | $43,800 | $2,400 | - | $12,700 | $58,900 |
| Wilma | - | $12,840 | - | - | - | - | $12,840 |
| Gustav | $600 | $600 | $12,600 | $600 | - | - | $14,400 |
| Ike | - | - | $12,000 | - | - | $12,800 | $24,800 |
| Total | $21,900 | $20,640 | $113,300 | $117,200 | $400 | $25,500 | $298,940 |

Source: USDA, NRCS, November 29, 2012.

## *Forest Service*[29]

The Forest Service (FS) administers programs for protecting and managing the natural resources of the National Forest System (primarily national forests and national grasslands) and for assisting states and non-industrial private forestland owners in protecting and managing the natural resources of non-federal forestlands. The FS provides financial and technical assistance, typically through state forestry agencies, to non-federal landowners to restore forests damaged by hurricanes (and other disasters). The state agencies are authorized to use such funds in numerous ways, such as assisting landowners to clear damaged trees and to plant new stands on cleared sites. While emergency and supplemental funding is sometimes enacted for natural disasters (e.g., hurricanes), the funding often is expended through ongoing, existing programs, and commonly cannot be distinguished from regular appropriations for these purposes (i.e., protecting and managing National Forest System lands and resources and assisting non-federal landowners in protecting and managing their forests). Funding for the FS to conduct work after a natural disaster can be categorized generally as recovery efforts and response efforts. Although the FS does not have the authority for specific programs to grant recovery assistance to states, the FS can use its regular program authorities to assist state and private landowners broadly following a disaster. For example, after a hurricane, the FS may receive supplemental funding under the state and private forestry (SPF) appropriation to conduct recovery work via a SPF program. Eight existing FS programs were used to assist the states following Hurricanes Katrina, Rita, Wilma, Gustav, and Ike (see **Table 6**).[30] The FS may also grant funding for the FSA Emergency Forest Restoration Program.[31] FS recovery funding amounts by state for the 2005 hurricanes (Katrina, Rita, and Wilma) and 2008 hurricanes (Gustav and Ike) are provided in **Table 7**.

Response tasks are identified through the National Response Framework (NRF), administered by FEMA, which grants the FS certain responsibilities (e.g., firefighting) to coordinate during a presidential-declared emergency or major disaster.[32] FS response funding by state for the 2005 hurricanes was not provided by the FS. The FS reports it spent approximately $92.4 million, $2.0 million, and $2.6 million for Hurricanes Katrina, Rita, and Wilma, respectively on response efforts in FS region 8.[33] FS response funding by state is available for the 2008 hurricanes.[34] The FS estimates it spent a total of $2.5 million on response efforts for Hurricane Gustav ($1.4 million in Alabama, $0.9 million in Louisiana, $0.1 million in Mississippi, and $0.1 million in Texas). The FS reports it spent a total of $2.1 million on response forts for Hurricane Ike (all funding spent in Texas).

**Table 6. Forest Service Programs Used to Grant Assistance after Hurricanes in 2005 and 2008**

| Program | Purpose | Authority |
|---|---|---|
| Cooperative Forest Health Protection | Provides federal financial and technical assistance to states to facilitate their survey and monitoring of forest health conditions and for the protection of forests and trees on state and private lands from insects, disease causing agents, and invasive plants. | 16 U.S.C. §2104 |
| Economic Action Program | To assist communities and their leaders in improving the efficiency and marketing of natural resource-based industries and in diversifying rural community economic bases. | 7 U.S.C. §6611-6617 |
| Emergency Forestry Conservation Reserve Program (temporary) | To provide assistance to nonindustrial private forest landowners who experienced a loss of 35% or more in merchantable timber from the 2005 hurricanes (Hurricane Katrina et al.) | P.L. 109-148 Section 107 |

## Table 6. (Continued)

| Program | Purpose | Authority |
|---|---|---|
| Forest Stewardship | To improve timber production and environmental protection on nonfederal forest lands | 16 U.S.C. §2103a |
| Hazard Fuel Mitigation | To assist communities in reducing threats from wildfires | 16 U.S.C. §2106 |
| State Fire Assistance | Provides technical and financial assistance to state cooperators | 16 U.S.C. §2106 |
| Urban and Community Forestry | To expand knowledge and awareness of the value of urban trees and to encourage the maintenance and expansion of urban tree cover | 16 U.S.C. §2105 |
| Volunteer Fire Assistance | To provide federal financial, technical, and other assistance to state foresters and other appropriate officials to organize, train, and equip fire departments in rural areas and rural communities to prevent and suppress fires | 16 U.S.C. §2106 |

Source: Compiled by CRS.

## Table 7. Forest Service 2005 and 2008 Hurricane Recovery Funding (Dollars in Thousands)

| Hurricane Year | State | Program | Grant Amount | FS Reimbursed |
|---|---|---|---|---|
| 2005 | Alabama | Forest Stewardship | $474 | $474 |
| 2005 | Alabama | Cooperative Forest Health Protection | $90 | $90 |
| 2005 | Alabama | Economic Action/Rural Development | $45 | $45 |
| 2005 | Alabama | State Fire Assistance | $369 | $369 |
| 2005 | Alabama | Urban and Community Forestry | $297 | $255 |
| 2005 | Alabama | Volunteer Fire Assistance | $50 | $50 |
| | | Totals | $1,325 | $1,282 |
| 2005 | Florida | Urban and Community Forestry | $730 | $615 |
| 2005 | Louisiana | Economic Action/Rural Development | $110 | - |
| | | Urban and Community Forestry/ State Fire Assistance/ Forest Stewardship/ Cooperative | $8,100 | $7,971 |

Federal Disaster Assistance after Hurricanes Katrina ... 23

| Hurricane Year | State | Program | Grant Amount | FS Reimbursed |
|---|---|---|---|---|
| 2005 | Louisiana | Forest Health | | |
| 2005 | Louisiana | Volunteer Fire Assistance | $517 | $517 |
| | | Totals | $8,728 | $8,489 |
| 2005 | Mississippi | Economic Action/Rural Development | $160 | $160 |
| | | Urban and Community Forestry/ State Fire Assistance/ Forest Stewardship/ Cooperative | $11,851 | $11,519 |
| 2005 | Mississippi | Forest Health | | |
| 2005 | Mississippi | Volunteer Fire Assistance | $626 | $553 |
| | | Totals | $12,638 | $12,232 |
| 2005 | Texas | Economic Action/Rural Development | $84 | $83 |
| | | Urban and Community Forestry/ State Fire Assistance/ Forest Stewardship/ Cooperative | $4,923 | $4,679 |
| 2005 | Texas | Forest Health | | |
| | | Totals | $5,007 | $4,763 |
| 2008 | Texas | State Fire Assistance | $6,500 | $4,089 |
| 2008 | Texas | Urban and Community Forestry (Carryover) | $50 | $50 |
| | | Totals | $6,550 | $4,139 |

Source: Data provided by the U.S. Forest Service, November 30, 2012.
Note: Numbers have been rounded.

## *Rural Housing Service*

The Rural Housing Service (RHS) provides loan and grant assistance for single-family and multifamily housing. RHS also administers the Community Facilities loan and grant program to provide assistance to communities for health facilities, fire and police stations, and other essential community facilities. Following the hurricanes, RHS provided housing relief to residents of the affected areas through payment moratoriums of six months, a three-month moratorium on initiating foreclosures under the single family guaranteed homeownership loans, loan forgiveness, loan reamortization, and refinancing. In addition, RHS provided temporary rental assistance to displaced family farm labor housing tenants. Assistance was provided for single-family homeowners (e.g., Section 502 loans), multi-family housing owners (e.g., Section 504 loans), and rental housing assistance (Section 521). Under P.L. 109-234, total outlays for RHS programs for the 2005 hurricanes were $63 million (see **Table 2**).[35]

The Disaster Relief and Recovery Supplemental Appropriations Act of 2008 (P.L. 110-329) provided funding for activities under the Rural Development Mission Area for relief and recovery from natural disasters (including hurricanes) during 2008. The act specifically provided $38 million for activities of the Rural Housing Service for areas affected by Hurricanes Katrina and Rita.

*Rural Utilities Service*

The Rural Utilities Service (RUS) is responsible for administering electric, telecommunications, and water assistance programs that help finance the infrastructure necessary to improve the quality of life and promote economic development in rural areas. Hurricane relief included grants for rebuilding, repairing, or otherwise improving water and waste disposal systems in designated disaster areas. Increased technical assistance under the Circuit Rider program was also provided to rural water districts. With the approval of lenders, RUS also suspended preauthorized debit payments for water and waste disposal loan guarantees for six months. Under permanent authority of P.L. 92-419, total outlays for RUS programs for the 2005 hurricanes were $14 million (see **Table 2**).

## Department of Commerce

### *National Oceanic and Atmospheric Administration*[36]

The federal government may provide disaster relief to the fishing industry when there is a commercial fishery failure. A commercial fishery failure occurs when fishermen endure hardships resulting from fish population declines or other disruptions to the fishery. The Department of Commerce can provide disaster assistance under Sections 308(b) and 308(d) of the Interjurisdictional Fisheries Act (16 U.S.C. §4107), and 312(a) and 315 of the Magnuson-Stevens Fishery Conservation and Management Act (16 U.S.C. §1864a and §1864). Funding is usually distributed as grants to states or regional marine fisheries commissions by the National Oceanic and Atmospheric Administration (NOAA) of the Department of Commerce.

Since 2005, Congress has appropriated almost $260 million of hurricane disaster relief to the Gulf of Mexico fishing industry (see **Table 8**). Of this total, $213 million was appropriated for damages and disruptions caused by Hurricanes Katrina and Rita (P.L. 109-234 and P.L. 110-28). Assistance provided for the direct needs of fishermen and related businesses, and supported related fisheries programs such as oyster bed and fishery habitat restoration, cooperative research, product marketing, fishing gear studies, and seafood testing. Many of these activities such as habitat restoration are ongoing management priorities for these fisheries. For damage caused by Hurricanes Gustav and Ike, $47 million was appropriated to restore damaged oyster reefs, remove storm debris, and rebuild fishing infrastructure in Texas and Louisiana (P.L. 110-329). In addition, $85 million was provided to NOAA for scanning, mapping, and removing marine debris; repairing and reconstructing the NOAA Science Center; procuring a replacement emergency response aircraft and sensor package; and other activities (P.L. 109-234 and P.L. 110-28).

**Table 8. Disaster Relief Funding for Commercial Fisheries (Obligations as of December 2012; Dollars in Thousands)**

| Commercial Fishery Disaster Assistance | Alabama | Florida | Louisiana | Mississippi | Texas | Total |
|---|---|---|---|---|---|---|
| Total | $44,633 | $6,233 | $134,190 | $62,042 | $11,375 | $258,473 |

Source: Gulf States Marine Fisheries Commission, Emergency Disaster Recovery Program, available at http://www.gsmfc.org/#:links@1:content@2. Vicki Schwantes, Budget Analyst, NOAA Budget Office, personal communication, December 10, 2012.

Notes: The total does not add to $260 million because $1,527 thousand was allocated for program administration. The table does not include funding for NOAA programs.

### *Economic Development Administration Economic Adjustment Assistance*[37]

The Economic Development Administration (EDA) was created with the passage of the Public Works and Economic Development Act of 1965 (PWEDA), P.L. 89-136, (42 U.S.C. §3121, et. al.) to provide assistance to communities experiencing long-term economic distress or sudden economic dislocation. Among the programs administered by EDA is the Economic

Adjustment Assistance (EAA) program. The PWEDA (42 U.S.C. §3149(c)(2)) authorizes EDA to provide EAA funds for:

> disasters or emergencies, in areas with respect to which a major disaster or emergency has been declared under the Robert T. Stafford Disaster Relief and Emergency Assistance Act for post-disaster economic recovery.[38]

In addition to funding disaster-recovery efforts using Emergency Assistance Act (EAA) funds available under its regular appropriation, 42 U.S.C. §3233 authorizes the appropriation of such sums as are necessary to fund EAA disaster recovery activities authorized under 42 U.S.C. §3149(c)(2). Funds appropriated under 42 U.S.C. §3233 may be used to cover up to 100% of the cost of a project or activity authorized under 42 U.S.C. §3149(c) (2). Funds appropriated under a regular appropriations act may be used to cover only 50% of the cost of disaster recovery activities. However, the authorizing statute also grants EDA the authority to increase the federal share of a project's cost to 100%.

**Disaster Assistance Grants**

Presidentially declared disasters or emergencies are one of five specific qualifying events eligible for EAA funding assistance.[39] EAA grants are competitively awarded and may be used to help finance public facilities; public services (including job training and counseling) business development (including funding a revolving loan fund (RLF); planning; and technical assistance that support the creation or retention of private sector jobs. Regions submitting an application for EAA disaster assistance must demonstrate a clear connection between the proposed project and disaster recovery efforts. EAA disaster grants can cover 100% of a project's cost.

In order to qualify for assistance, the Secretary of Commerce must find that a proposed project or activity will help the area respond to a severe increase in unemployment, or economic adjustment problems resulting from severe changes in economic conditions. EAA regulations also require an area seeking such assistance to prepare or have in place a Comprehensive Economic Development Strategy (CEDS) outlining the nature and level of economic distress in the region, and proposed activities that could be undertaken to support private-sector job creation or retention efforts in the area.

**Funding Narrative**

Congress did not provide EAA supplemental appropriations for disaster recovery activities related to Hurricanes Katrina, Rita, or Wilma. However, EDA allocated $24.2 million from its regular appropriations in response to the hurricanes of 2005. In response to Hurricanes Gustav and Ike and other disasters occurring in 2008, Congress appropriated $400 million in EAA disaster supplemental funding when it approved P.L. 110-329. It also appropriated an additional $100 million in supplemental EAA disaster assistance without limiting it to disasters occurring in a specific year when it passed the Supplemental Appropriations Act of 2008, P.L. 110-252.

Of the $500 million appropriated for EAA disaster grants in 2008, EDA allocated, based on its 2010 annual report to Congress, the latest data available, a total of $63.8 million to 33 recipients in five of the six states identified in this report. This included $25.4 million to 11 projects in the state of Louisiana, $16.6 million to 9 projects in Florida, $14.3 million to 9 projects in Texas, $5.7 million to 3 projects in Mississippi, and $4.8 million for 1 project in Alabama.[40]

A summary, by state, of EAA disaster recovery grants for projects funded in response to Hurricanes Katrina, Rita, Wilma, Gustav, and Ike is presented in **Table 9**.

**Table 9. Distribution of EAA Disaster Recovery Grants for Selected States by Hurricane Event**
**(Dollars in Thousands)**

| Hurricane | Alabama | Florida | Louisiana | Mississippi | Tennessee | Texas | Total |
|---|---|---|---|---|---|---|---|
| Katrina | - | - | - | - | - | - | - |
| Rita | - | - | - | -- | - | - | - |
| Wilma | - | - | - | -- | - | - | - |
| Gustav | - | - | - | -- | - | - | - |
| Ike | $4,758 | $16,623 | $25,400 | $5,745- | - | $14,301 | $63,827 |
| Total | $4,758 | $16,623 | $25,400 | $5,745- | - | $14,301 | $63,827 |

Source: U.S. Economic Development Administration, *Fiscal Year 2010 Annual Report*.
Notes: Table does not include funds awarded to states using the program's regular appropriations. Congress did not provide supplemental EAA disaster appropriations for Hurricanes Katrina, Wilma, and Rita.

## Department of Defense (Civil)[41]

### Army Corps of Engineers

**Civil Works Program**

The U.S. Army Corps of Engineers (Corps) is a unique federal agency in the Department of Defense, with military and civilian responsibilities. Under its civil works program, the Corps plans, builds, operates, and maintains a wide range of water resources facilities, including hurricane protection and flood damage reduction projects, and performs emergency actions for flood and coastal emergencies.

**Table 10** shows, for each Gulf Coast state, the direct appropriations that the Corps received for its water resources work related to the five hurricanes. According to data the Corps provided to CRS, of the total $15.6 billion appropriated, more than $11.2 billion has been obligated.

**Table 10. Disaster Relief Funding Appropriations for the Army Corps of Engineers (Dollars in Thousands)**

| Army Corps of Engineers | Alabama | Florida | Louisiana | Mississippi | Texas | Total |
|---|---|---|---|---|---|---|
| Civil Works Appropriations | $3,000 | $57,000 | $14,768,000 | $558,000 | $207,000 | $15,593,000 |

Source: CRS correspondence with Army Corps of Engineers Budget Office, 2012.

## Department of Defense (Military)[42]

### Military Personnel

The Military Personnel accounts fund military pay and allowances, permanent change of station travel, retirement and health benefit accruals, uniforms, and other personnel costs. For the hurricane response efforts, funds have been used primarily to pay per diem to DOD personnel evacuated from affected areas, for the pay and allowances of activated Guard and Reserve personnel supporting the hurricane relief effort, and for increased housing allowances to compensate for housing rate increases in hurricane-affected areas. Military personnel funds obligated by the Alabama, Florida, Texas,

Louisiana, and Mississippi National Guard are detailed in **Table 11**. Data on the obligation of other Military Personnel funds, by state, were not readily available.

### *Operations and Maintenance*

The Operations and Maintenance (O&M) accounts fund training and operation costs, pay for civilians, maintenance service contracts, fuel, supplies, repair parts, and other expenses. For the hurricane response efforts, funds have been used primarily to repair facilities, establish alternate operating sites for displaced military organizations, repair and replace equipment, remove debris, clean up hazardous waste, repair utilities, evacuate DOD personnel from affected areas, and support the operations of activated Army and Air National Guard units. O&M funds obligated by the Alabama, Florida, Texas, Louisiana, and Mississippi National Guard are detailed in **Table 11**. Data on the obligation of other O&M funds, by state, were not readily available.

### *Procurement*

The Procurement accounts generally fund the acquisition of aircraft, ships, combat vehicles, satellites, weapons, ammunition, and other capital equipment. For the hurricane response efforts, $2.85 billion was appropriated, of which $2.5 billion was used primarily to pay for extraordinary shipbuilding and ship repair costs, including not only damage to ships under construction and replacement of equipment and materials, but also additional overhead and labor costs resulting from schedule delays due to the hurricane damage to shipyards, primarily Avondale in New Orleans, Louisiana, and Ingalls in Pascagoula, Mississippi.[43] These funds also included $140 million to improve the infrastructure at damaged shipyards.[44] Budget authority, obligations, and outlays for procurement, allocated by state for Alabama, Florida, Texas, Louisiana, and Mississippi are detailed in **Table 11**.

### *Research, Development, Test, and Evaluation*

The Research, Development, Test, and Evaluation (RDT&E) accounts fund modernization efforts by way of basic and applied research, creation of technology-demonstration devices, developing prototypes, and other related costs. For the hurricane response efforts, funds have been used to replace damaged test equipment and repair damaged test facilities. Data allocating RDT&E funds by state were not readily available.

## Military Construction (MILCON) and Family Housing

The MILCON accounts fund the acquisition, construction, installation, and equipment of temporary or permanent public works, military installations, facilities, and real property.

The Family Housing Construction accounts fund costs associated with the construction of military family housing (including acquisition, replacement, addition, expansion, extension, and alteration), while the Family Housing O&M accounts fund expenses such as debt payment, leasing, minor construction, principal and interest charges, and insurance premiums on military family housing.

For the hurricane response efforts, $1.4 billion was appropriated to finance the planning, design, and construction of military facilities and infrastructure that were damaged or destroyed by hurricane winds and water. Of this, $918 million was dedicated to military operations and training facilities, while an additional $460 million was appropriated for family housing construction and family housing O&M to rebuild destroyed, damaged, or new housing units and a housing office.

Budget authority for MILCON and family housing construction allocated to the states of Alabama, Florida, Texas, Louisiana, and Mississippi is detailed in Table 11. Of the $1.4 billion appropriated, $1.2 billion could be allocated to the five specified states, while $167 million was devoted to planning and design activities not associated with specific locations.

## Management Funds

This category includes the Defense Working Capital Fund, the National Defense Sealift Fund, and a commissary fund. For the hurricane response efforts, these funds have been used primarily to rebuild and repair damaged commissaries, replace commissary inventories, and cover transportation and contingency costs of the Defense Logistics Agency. Data allocating these funds by state were not readily available.

## Other Department of Defense Programs

This category includes the Defense Health Program (DHP) and the Office of the Inspector General (OIG). The DHP title funds medical and dental care to current and retired members of the Armed Forces, their family members, and other eligible beneficiaries.

For the hurricane response efforts, these funds have been used primarily to pay for costs associated with displaced beneficiaries seeking care from private-sector providers rather than at military health care facilities, to pay the health care costs of activated Guard and Reserve personnel, and to replace medical supplies and equipment.

Data allocating DHP funds by state were not readily available. Of the $589,000 appropriated for the OIG, $263,000 was provided to replace and repair damaged equipment in the Inspector General's office in Slidell, Louisiana, and to cover relocation costs.

**Table 11. Disaster Relief Funding by the Department of Defense (Military) (Budget Authority/Obligations/Outlays, as of December, 2012; Dollars in Thousands)**

| Name of Program | Alabama | Florida | Louisiana | Mississippi | Texas | Total |
|---|---|---|---|---|---|---|
| Military Personnel (National Guard only)[a] | $7,192 | $4,091 | $126,982 | $27,123 | $15,974 | $181,361 |
| Operations and Maintenance (National Guard only)[b] | $1,407 | $1,759 | $89,538 | $112,721 | $12,440 | $217,866 |
| Procurement: Budget Authority[c] | $60,048 | - | $770,647 | $1,698,581 | - | $2,529,277 |
| Obligations | $60,007 | - | $770,546 | $1,698,193 | - | $2,528,746 |
| Outlays | $54,996 | - | $697,584 | $1,567,619 | - | $2,320,199 |
| Military Construction and Family Housing[d] | - | - | $371 | $840 | - | $1,378 |

Sources: The National Guard Personnel and O&M figures are CRS calculations based on data provided by the National Guard Bureau. Procurement figures were provided by the Navy. Military Construction and Family Housing figures are CRS calculations based on data contained in the conference committee reports that accompanied the relevant appropriations acts.

[a] National Guard figures are expressed in terms of obligations.
[b] The obligated funds for National Guard personnel and O&M were for hurricane response purposes in the specified states from 2005-2012, but they may not correspond in all cases to the emergency funds appropriated by Congress for hurricane relief purposes specified in Table 1 of this Report. An indeterminate amount of the funding came from regular appropriated funds.
[c] Procurement figures are expressed in terms of budget authority, obligations, and outlays; budget authority is nearly identical to obligations.
[d] Military construction figures are expressed in terms of budget authority; $167 million is not geographically specific.

# Department of Education[45]

## Elementary and Secondary Education

### Program Authorities[46]

Following the Gulf Coast hurricanes, funding to support elementary and secondary schools affected by Hurricane Katrina or Hurricane Rita was provided through three public laws: P.L. 109-148 ($1.4 billion), P.L. 109-234 ($235 million), and P.L. 110-28 ($30 million).

- P.L. 109-148 created two new programs: (1) Immediate Aid to Restart School Operations ($750 million) and (2) Temporary Impact Aid for Displaced Students ($645 million) specifically designed to address needs resulting from the hurricanes and to provide support to local educational agencies (LEAs) through an existing federal education program administered by the U.S. Department of Education (ED).[47] It also added $5 million to the McKinney-Vento Homeless Assistance Act to serve homeless children and youth who had been displaced by the Gulf Coast hurricanes.
- P.L. 109-234 provided additional funding of $235 million for the Temporary Impact Aid for Displaced Students enacted under P.L. 109-148.
- P.L. 110-28 appropriated $30 million for elementary and secondary schools affected by the hurricanes through the Hurricane Educator Assistance program to assist in recruiting, retaining, and compensating staff in those schools.

Congress then appropriated an additional $15 million through P.L. 110-329 to provide support to LEAs whose enrollment of homeless students increased as a result of hurricanes, including Hurricanes Gustav and Ike, floods, or other natural disaster during 2008. Most recently, Congress appropriated $12 million through P.L. 111-117 for the Gulf Coast Recovery Initiative to improve education in areas affected by Hurricanes Katrina, Rita, or Gustav. A brief description of each of these programs and the amount of funding each received is presented below. **Table 12** details how much funding various states received under each of the programs.

### Immediate Aid to Restart School Operations

The Immediate Aid to Restart School Operations provided support for LEAs and non-public schools in Louisiana, Mississippi, Alabama, and Texas to restart school operations, reopen schools, and re-enroll students. P.L. 109-148 provided $750 million for this program. This program is no longer authorized.

### Temporary Emergency Impact Aid for Displaced Students

The Temporary Emergency Impact Aid for Displaced Students program provided federal funding to assist schools in enrolling students who had been displaced by the Gulf Coast hurricanes. Funds were made available to LEAs and schools based on the number of displaced students that enrolled, irrespective of whether the school in which parents chose to enroll their child was a public or non-public school. P.L. 109-148 appropriated $645 million for this program. Subsequently, P.L. 109-234 appropriated an additional $235 million for this program, bringing the total program appropriation to $880 million.[48] Portions of the funds appropriated were provided to 49 states[49] and the District of Columbia based on the number of displaced students each enrolled. Louisiana, Texas, and Mississippi received the largest proportion of funds. This program is no longer authorized.

### Hurricane Educator Assistance Program

The Hurricane Educator Assistance Program made federal funding available to Louisiana, Mississippi, and Alabama to use for recruiting, retaining, and compensating school staff who committed to work for at least three years in public elementary and secondary schools affected by Hurricanes Katrina or Rita. States were required to apply to receive funds, and the funds were allocated based on the number of public elementary and secondary schools that were closed for 19 days or more from August 29, 2005, through December 31, 2005. P.L. 110-28 provided $30 million for these purposes. Funds were provided to Louisiana and Mississippi only. This program is no longer authorized.

### McKinney-Vento Homeless Assistance Act

The McKinney-Vento Homeless Assistance Act provides funding to states to ensure that homeless children and youth are provided equal access to a free, appropriate public education in the same manner as provided other children and youth.[50] P.L. 109-148 appropriated $5 million for this program for LEAs serving homeless children and youth who had been displaced by Hurricane

Katrina or Hurricane Rita. Eight states received funding under this program, with the largest grants provided to Texas and Louisiana.[51] While the McKinney-Vento Homeless Assistance Act continues to provide funding related to the education of homeless students, the provisions enacted specifically in response to the Gulf Coast hurricanes are no longer authorized.

### Homeless Education Disaster Assistance[52]

P.L. 110-329 provided $15 million to LEAs whose enrollment of homeless students increased as a result of hurricanes, floods, or other natural disasters that occurred during 2008 and for which the President declared a major disaster under Title IV of the Stafford Act. ED was required to distribute the funds through the McKinney-Vento Homeless Assistance Act based on demonstrated need. These funds provided assistance to LEAs in Gulf Coast states affected by Hurricanes Gustav and Ike, as well as LEAs affected by natural disasters in other parts of the nation, such as flooding in the Midwest. The majority of the funds were provided to LEAs in Louisiana and Texas.[53] This program is no longer authorized.

### Gulf Coast Recovery Initiative

P.L. 111-117 provided $12 million for competitive awards to LEAs located in counties in Louisiana, Mississippi, and Texas that were designated by FEMA as counties eligible for individual assistance as a result of damage caused by Hurricanes Katrina, Rita, or Gustav. The funds had to be used to improve education in areas affected by these hurricanes and had to be used for activities such as replacing instructional materials and equipment; paying teacher incentives; modernizing, renovating, or repairing school buildings; supporting charter school expansion; and supporting extended learning time activities. The majority of the funds were provided to LEAs in Louisiana. This program is no longer authorized.

### Higher Education Program Authorities

Appropriations to support institutions of higher education (IHEs) following the Gulf Coast hurricanes of 2005 were provided through P.L. 109-148 ($200 million), P.L. 109-234 ($50 million), and P.L. 110-28 ($30 million). P.L. 110-329 subsequently provided another $15 million for IHEs in areas affected by hurricanes, including Hurricanes Gustav and Ike, floods, or other natural disasters in 2008. **Table 12** details the amount of funding allocated to various states under these provisions.

## Hurricane Education Recovery

Of the $200 million provided under P.L. 109-148 for higher education, $95 million was specifically appropriated for the Louisiana Board of Regents, and $95 million was specifically appropriated for the Mississippi Institutes of Higher Learning for hurricane education recovery from the 2005 Gulf Coast hurricanes. Subsequently, P.L. 109-234 and P.L. 110-28 provided additional funds for hurricane education recovery under the Fund for the Improvement of Postsecondary Education (FIPSE), authorized by Title VII of the Higher Education Act, to assist IHEs adversely affected by the 2005 Gulf Coast hurricanes. Under both laws, funds were provided to help defray the expenses incurred by IHEs that were forced to close, relocate, or reduce their activities due to hurricane damage. Under P.L. 110-28, IHEs also were permitted to use these funds to make grants to students enrolled at these institutions on or after July 1, 2006. A total of $80 million was provided for IHEs affected by Hurricane Katrina or Hurricane Rita under FIPSE for hurricane education recovery. The majority of funds appropriated for hurricane education recovery were provided to Mississippi and Louisiana. These activities are no longer authorized.

## Funds to Assist IHEs Enrolling Displaced Students

The remaining $10 million appropriated under P.L. 109-148 for higher education disaster relief was provided to assist IHEs with unanticipated costs associated with the enrollment of students displaced as a result of Hurricane Katrina or Hurricane Rita. Overall, 99 IHEs in 24 states and the District of Columbia received funds related to the enrollment of displaced higher education students.[54] Louisiana and Texas received the largest state grants. This program is no longer authorized.

## Higher Education Disaster Relief[55]

P.L. 110-329 provided an additional $15 million[56] for IHEs that were located in an area affected by hurricanes, floods, and other natural disasters that occurred during 2008 and for which the President declared a major disaster under Title IV of the Stafford Act. Funds provided through the Higher Education Disaster Relief program could be used to defray the expenses incurred by IHEs that were forced to close or relocate or whose operations were adversely affected by the natural disaster, and to provide grants to students who attended such IHEs for academic years beginning on or after July 1, 2008. The majority of these funds were provided to Louisiana and Texas for hurricane-related education disaster assistance related to Hurricanes Gustav and Ike.[57] This program is no longer authorized.

**Funding Summary**

Following the Gulf Coast hurricanes of 2005, Congress appropriated $1.943 billion for ED to provide support to LEAs, schools, and IHEs in the Gulf Coast region and nationwide that were affected by Hurricane Katrina or Hurricane Rita.[58] Subsequently, FY2009 supplemental appropriations provided an additional $30 million for education-related disaster relief for LEAs and IHEs affected by natural disasters during the 2008 calendar year. Most recently, FY2010 appropriations provided an additional $12 million for LEAs located in specific areas affected by Hurricanes Katrina, Rita, or Gustav. Of the $1.985 billion provided for education-related disaster relief and administered by ED since the Gulf Coast hurricanes, nearly all of these funds (92%) were provided to Alabama, Florida, Louisiana, Mississippi, Tennessee, and Texas in response to Hurricanes Katrina, Rita, Gustav, and Ike. **Table 12** details how much of this funding was allocated to each of these states for each of the programs discussed in this section.

**Table 12. Disaster Relief Funding Administered by the Department of Education Provided in Response to Hurricanes Katrina, Rita, Gustav, and Ike (Dollars in Thousands (cumulative obligations))**

| Department of Education | Alabama | Florida | Louisiana | Mississippi | Tennessee | Texas | Total |
|---|---|---|---|---|---|---|---|
| Elementary and secondary education ||||||||
| Immediate Aid to Restart School Operations | $3,750 | – | $445,604 | $222,493 | – | $78,153 | $750,000 |
| Emergency Impact Aid for Displaced Students[a] | $36,605 | $27,214 | $291,717 | $100,787 | $19,001 | $250,890 | $726,213 |
| McKinney-Vento Homeless Education Assistance Program | $247 | $196 | $1,564 | $687 | $122 | $1,687 | $4,504 |

| Department of Education | Alabama | Florida | Louisiana | Mississippi | Tennessee | Texas | Total |
|---|---|---|---|---|---|---|---|
| Hurricane Educator Assistance Program | – | – | $22,593 | $7,407 | – | – | $30,000 |
| Homeless Education Disaster Assistance[b] | – | – | $1,171 | – | – | $12,256 | $13,427 |
| Gulf Coast Recovery Initiative | – | – | $8,624 | $2,638 | – | $739 | $12,000 |
| Subtotal for elementary and secondary education | $40,602 | $27,410 | $771,273 | $334,012 | $19,123 | $343,724 | $1,536,144 |
| Higher education | | | | | | | |
| Hurricane Education Recovery | $301 | $1,507 | $145,663 | $117,878 | – | $4,651 | $270,000 |
| Funds to Assist Institutions of | $357 | $34 | $5,748 | $327 | $95 | $1,750 | $8,312 |
| Higher Education Enrolling Displaced Students | | | | | | | |
| Higher Education Disaster Relief Program[c] | – | – | $3,524 | – | – | $8,067 | $11,591 |
| Subtotal postsecondary education | $658 | $1,541 | $154,935 | $118,206 | $95 | $14,468 | $289,903 |
| Total | $41,261 | $28,952 | $926,208 | $452,217 | $19,218 | $358,192 | $1,826,046 |

Source: Table prepared by CRS, December 11, 2012, based on published and unpublished data available from the U.S. Department of Education (ED).

Notes: Details may not add to totals due to rounding.

[a] Under the Emergency Impact Aid program, $1.9 million of the $880 million appropriated was not allocated to states, as the funds were not needed. Thus, the total appropriated amount is higher than the amount allocated and shown on the table.

[b] None of these funds were provided in response to the Gulf Coast hurricanes of 2005. While data were not available from ED on the specific disasters experienced by the LEAs that received funding, data were available on the specific types of disasters for which IHEs received funds under P.L. 110-329. According to these data, all IHEs in Louisiana that received funds were affected by Hurricane Gustav or Ike. Most IHEs in Texas that received funds were affected by Hurricane Ike. A few IHEs in Texas were affected by Hurricane Dolly, accounting for a relatively small portion of the funds allocated to IHEs in Texas. IHEs in Florida that received funding were affected by Tropical Storm Fay. Thus, all funds provided to LEAs in Louisiana and Texas were included in the table, while funds provided to LEAs in Florida were not included.

[c] None of these funds were provided in response to the Gulf Coast hurricanes of 2005. Only funds provided to IHEs in response to Hurricane Gustav or Hurricane Ike were included in the table.

## Department of Health and Human Services

### *Administration for Children and Families*[59]

### Head Start

The federal Head Start program, authorized at 42 U.S.C. §9801 et seq., provides comprehensive early childhood development services to low-income children.[60] The program seeks to promote school readiness by enhancing the social and cognitive development of children through the provision of educational, health, nutritional, social, and other services. Federal Head Start funds are provided directly to local grantees (e.g., public and private nonprofit and for-profit agencies) rather than through states. Most children served in Head Start programs are three- and four-year olds, but services are authorized for children from birth through compulsory school age.

In December 2005, Congress appropriated $90 million in supplemental Head Start funds for the costs of serving displaced children and the renovation of Head Start facilities affected by the Gulf Coast hurricanes of 2005.[61] The Department of Health and Human Services (HHS) Administration for Children and Families (ACF) reported awarding approximately $74 million of the total appropriation based on grantee requests; the remaining funds ($16 million) reverted to the U.S. Treasury Department.[62] The majority of the funds awarded to grantees ($72.5 million, or 98% of the $74 million) went to Head Start programs in Alabama, Florida, Louisiana, Mississippi, and Texas (see **Table 13**).

## Social Services Block Grant

The Social Services Block Grant (SSBG), permanently authorized by 42 U.S.C. §1397 et seq., is a flexible source of funds that states use to support a wide variety of social services activities, ranging from child care to special services for the disabled.[63] States have broad discretion over the use of SSBG funds, which are typically allocated to states according to population-based formula.

In December 2005, Congress appropriated $550 million in supplemental SSBG funds for necessary expenses related to the consequences of the Gulf Coast hurricanes of 2005.[64] ACF distributed these funds based on the number of FEMA registrants from Hurricanes Katrina, Rita, and Wilma, as well as the percent of individuals in poverty in each state. Funds were allocated to all states that took in evacuees, not just the states that were directly affected. The appropriations language expanded potential services for which these funds could be used to include "health services (including mental health services) and for repair, renovation, and construction of health facilities (including mental health facilities)."

In September 2008, Congress appropriated $600 million for necessary expenses resulting from major disasters occurring in 2008, including hurricanes, floods, and other natural disasters, as well as expenses resulting from Hurricanes Katrina and Rita.[65] ACF reserved a portion of these funds for states affected by major disasters of 2008 and a portion for states facing ongoing needs as a result of Hurricanes Katrina and Rita.[66] ACF distributed both sets of funds based on each state's share of FEMA registrants, as well as the overall population for each state. Like the previous supplemental, the 2008 supplemental appropriation again expanded potential services for which SSBG funds could be used, this time to include "health services (including mental health services) and for repair, renovation, and construction of health facilities (including mental health facilities), child care centers, and other social services facilities."

Combined, these two supplemental appropriations provided $1.150 billion for the SSBG. According to ACF, the bulk of these funds—$944 million, or 82% of the $1.150 billion—were allocated to Alabama, Florida, Louisiana, Mississippi, and Texas (see **Table 13**).[67]

Typically, SSBG funds are subject to a two-year expenditure period—meaning that funds must be spent by the end of the fiscal year subsequent to the fiscal year in which they were allotted to states.[68] However, most states had not spent all of their funds from either supplemental within the standard two-year period and, in both cases, Congress passed legislation extending the spending deadline for these supplemental funds.[69] According to data from ACF, states had spent about $521 million (95%) of the $550 million supplemental before the extended deadline of September 30, 2009. ACF data suggest that states had spent more than $522 million (87%) of the $600 million supplemental before the extended expenditure deadline of September 30, 2011.[70] Unspent funds were to revert to the U.S. Treasury.

According to the FY2009 SSBG annual report (most recent available), states spent supplemental funds on 28 of the 29 SSBG service categories defined in federal regulation,[71] including education and training, counseling services, and health-related services.[72] The FY2009 report indicated that most supplemental funds were spent in the "other services" category, including expenditures for certain construction and renovation costs, as well as costs related to certain health and mental health services. Notably, the FY2009 annual report only includes expenditures from the December 2005 supplemental appropriation. Forthcoming annual reports will presumably also include data on the later supplemental.

**Table 13. Disaster Relief Funding for Programs at the HHS Administration for Children and Families (Cumulative Allocations as of July 2010; Dollars in Thousands)**

| HHS Administration for Children and Families | Alabama | Florida | Louisiana | Mississippi | Texas | Total |
|---|---|---|---|---|---|---|
| Head Start | $1,390 | $114 | $44,995 | $22,212 | $3,796 | $72,507 |
| Social Services Block Grant (SSBG) | $40,945 | $89,194 | $350,639 | $156,535 | $306,805 | $944,117 |
| Total | $42,335 | $89,308 | $395,634 | $178,747 | $310,601 | $1,016,624 |

Source: CRS interpretation based on data from the HHS Administration for Children and Families (ACF). Head Start data are from ACF's FY2008 Justification of Estimates for Appropriations Committees. SSBG data are for combined supplemental allocations, based on data available at http://www.acf.hhs.gov/programs/ocs/resource/ supplemental (for the 2005 supplemental) and http://www.acf.hhs.gov/programs/ocs/resource/grant-awards (for the 2008 supplemental).

Notes: Totals shown for the SSBG reflect a combination of supplemental funds appropriated by P.L. 109-148 in December 2005 and P.L. 110-329 in September 2008. Notably, the 2008 SSBG supplemental was appropriated for expenses resulting from major disasters occurring during 2008, as well as Hurricanes Katrina and Rita. Thus, the allocations shown in this table include some funds that were allocated for disasters other than Hurricanes Katrina, Rita, Wilma, Gustav, and Ike (e.g., Tropical Storm Fay and Hurricane Dolly).

Federal Disaster Assistance after Hurricanes Katrina ... 41

## *Public Health and Medical Assistance*[73]

### DRF-Funded Mission Assignments

The Department of Health and Human Services (HHS) is the coordinating agency for Emergency Support Function 8 (ESF #8), Public Health and Medical Services, under the *National Response Framework*.[74] The Stafford Act authorizes reimbursements to HHS for many of its emergency or major disaster response activities, including (among others): deployment of operational assets (medical surge and mortuary teams, portable field hospitals, and the Strategic National Stockpile of drugs and medical supplies); disease surveillance; food and water safety activities; and workforce assistance to health departments.[75] Reimbursements to HHS for mission assignments are presented in **Table 18**, **Table 19**, and **Table 20**.

### DRF-Funded Crisis Counseling Program (CCP)

Pursuant to Section 416 of the Stafford Act, the President may provide assistance for the establishment of crisis counseling services in areas affected by declared major disasters. CCP, a program to provide short-term mental health screening, counseling, and referral services in presidentially declared disasters, is jointly administered by FEMA, the Substance Abuse and Mental Health Services Administration (SAMHSA) in HHS, and affected states. Amounts provided to each state for the response to the Gulf Coast hurricanes are displayed in **Table 14**.

### Federal Assistance for Health Care

In response to Hurricane Katrina, Congress authorized and appropriated a one-time program of up to $2.1 billion to cover full federal funding of the state match that would normally have been required under the Medicaid and State Children's Health Insurance (CHIP) programs, and the costs of uncompensated care, for eligible individuals from disaster-affected areas. Assistance was provided both to directly affected states and to certain states that hosted evacuees. Funding was also authorized "to restore access to health care in impacted communities," and was provided to stabilize the primary care workforce in three directly affected states: Alabama, Louisiana, and Mississippi.[76] Outlay amounts are presented in **Table 15**.[77]

### Appropriations to Existing HHS Accounts

In response to the 2005 hurricanes, Congress provided, in emergency supplemental appropriations for affected areas, $4 million for communications equipment for community health centers, and $8 million for mosquito abatement in affected states.[78] The amounts obligated from this emergency supplemental funding are presented in **Table 16**.

**Grants from Existing HHS Accounts**

In some cases, funds available in existing HHS accounts were provided for hurricane relief. For example, the Centers for Medicare and Medicaid Services (CMS) Emergency Prescription Assistance Program provided up to $2 million in individual assistance for affected counties in Texas following Hurricane Ike. Also, the HHS Office of Minority Health provided $12 million in grants to minority-serving organizations following Hurricane Katrina. Third, SAMHSA Emergency Response Grants (SERG) provided funds to states for mental health and substance abuse services following Hurricane Katrina.[79] Amounts for SERG grants are presented in **Table 13**.

*Administrative Waivers*

The federal government funds a significant portion of the nation's health care costs, through the Medicare and Medicaid programs, veterans and Indian health care systems, and other activities.

In response to the major hurricanes, HHS invoked numerous waiver authorities that allowed state, local, tribal, and private health care providers and facilities affected by the disasters to continue receiving federal health care services and/or reimbursements under altered conditions, such as the use of temporary facilities, the use of volunteer providers, and care provided to individuals not usually eligible.[80] Although these waivers did not provide new funds to disaster-affected areas, they prevented the loss of substantial federal revenues. Several HHS agencies also allowed states to reprogram federal grant funds, including from most of the grants administered by the Centers for Disease Control and Prevention (CDC).

*Public Health Emergency Fund*

The Secretary of HHS has authority to use a no-year fund[81] for public health emergencies. However, the fund has not had a balance since the 1990s, so it was not available for the response to the 2005 and 2008 hurricanes.[82]

**Table 14. Disaster Relief Funding for Crisis Counseling, Mental Health, and Substance Abuse Services (Allocations as of June 2010; Dollars in Thousands)**

| Hurricane | Alabama CCP | Alabama SERG | Florida CCP | Florida SERG | Louisiana CCP | Louisiana SERG | Mississippi CCP | Mississippi SERG | Texas CCP | Texas SERG | Total by Program CCP | Total by Program SERG |
|---|---|---|---|---|---|---|---|---|---|---|---|---|
| Katrina | $3,019 | $100 | - | - | $100,436 | $200 | $41,101 | $150 | - | $150 | $144,556 | $600 |
| Rita | - | - | - | - | $4,484 | - | - | - | $2,709 | - | $7,193 | - |
| Wilma | - | - | $10,401 | - | - | - | - | - | - | - | $10,401 | - |
| Gustav | - | - | - | - | $16,476 | - | - | - | - | - | $16,476 | - |
| Ike | - | - | - | - | - | - | - | - | $8,267 | - | $8,267 | - |
| Total | $3,019 | $100 | $10,401 | - | -$121,396 | $200 | $41,101 | $150 | $10,976 | $150 | $186,893 | $600 |

Source: Information for Hurricanes Katrina, Rita, and Wilma from FEMA, "Disaster Relief Fund: Monthly Status Report," (FY2010 Report to Congress), June 22, 2010, pp. 11-14; and HHS, "HHS Awards $600,000 in Emergency Mental Health Grants to Four States Devastated by Hurricane Katrina," press release, September 13, 2005, available at http://www.hhs.gov/news/. Information for Hurricanes Gustav and Ike provided by FEMA Office of External Affairs, July 14, 2010.

Notes: CCP is the Crisis Counseling Program. SERG is SAMHSA Emergency Response Grants. A hyphen indicates that no funds were provided. Although CCP allocations may have continued since June 2010, FEMA has not provided incident-specific funding information since that time. The SERG allocations as presented are final.

### Table 15. Disaster Relief Funding for Health Care Costs and Infrastructure
(Outlays as of December 31, 2012; Dollars in Thousands)

| Source | Alabama | Florida | Louisiana | Mississippi | Texas |
|---|---|---|---|---|---|
| Health care costs | $240,300 | $1,800 | $741,100 | $581,400 | $33,100 |
| Primary care stabilization | $38,300 | - | $57,600 | $92,800 | - |
| Total | $278,600 | $1,800 | $998,700 | $674,200 | $33,100 |

Source: HHS Office of the Assistant Secretary for Financial Resources, February 26, 2013.
Notes: Authority and appropriations pursuant to the Deficit Reduction Act (DRA), Section 6201, limited to the Hurricane Katrina response. Amounts included $2.0 billion appropriated under DRA, and authority to transfer up to $100 million previously appropriated to the National Disaster Medical System (NDMS). For health care costs only, funding was provided to 21 additional states and the District of Columbia, which hosted evacuees.

### Table 16. Disaster Relief Funding for Communications Equipment and Mosquito Abatement
(Obligations as of July 2009; Dollars in Thousands)

| Purpose | Alabama | Florida | Louisiana | Mississippi | Texas | Total |
|---|---|---|---|---|---|---|
| Communications equipment | $667 | $667 | $667 | $667 | $663 | $3,331 |
| Mosquito abatement | $798 | - | $3,202 | $2,871 | $1,109 | $7,980 |
| Total | $1,465 | $667 | $3,869 | $3,538 | $1,772 | $11,311 |

Source: HHS: Health Resources and Services Administration (HRSA) Office of Legislation; and CDC Washington Office, July 15, 2009.
Notes: Assistance provided for the response to Hurricane Katrina pursuant to P.L. 109-234. North Carolina also received a comparable award for communications equipment. On July 15, 2010, the HHS Office of the Assistant Secretary for Financial Resources confirmed that the amounts appropriated–$4 million for communications equipment and $8 million for mosquito abatement–had been fully obligated.

# Department of Homeland Security[83]

*Federal Emergency Management Agency*

**Authority**

The Stafford Act authorizes the President to issue major disaster or emergency declarations in response to incidents in the United States that overwhelm state and local governments.[84] Section 403(a)(1) of Stafford authorizes the President to direct federal resources to provide assistance essential to meeting immediate threats to life and property resulting from a major disaster.[85] Section 304 of the Stafford Act authorizes the reimbursement of other agencies from funds appropriated to the DRF for services or supplies furnished under the authority of the Stafford Act.[86]

**Program Description**

The primary mission of FEMA is to "reduce the loss of life and property and protect the Nation from all hazards, including natural disasters, acts of terrorism, and other man-made disasters, by leading and supporting the Nation in a risk-based, comprehensive emergency management system of preparedness, protection, response, recovery, and mitigation."[87]

FEMA provides assistance to states, local governments, tribal nations, individuals and families, and certain nonprofit organizations through the Disaster Relief Fund (DRF).[88] The more significant aid programs authorized under the Stafford Act include the Public Assistance Program (PA);[89] and the Individual and Household Program (IHP), which includes Other Needs Assistance (ONA)[90] and Debris Removal,[91] the Hazard Mitigation Grant Program (HMGP),[92] and Essential Assistance.[93]

P.L. 112-175[94] requires the FEMA Administrator to provide a report by the fifth day of each month on the DRF which includes DRF funding summaries. The DRF report provides funding information by state for the 2005 and 2008 hurricanes. As shown in **Table 17**, the DRF report aggregates funding for Hurricanes Katrina, Rita, and Wilma.

**Table 17. Disaster Relief Funding by the Federal Emergency Management Agency for Hurricanes Katrina, Rita, Wilma, Gustav, and Ike (Cumulative Obligations as of February 5, 2013; Dollars in Millions)**

| Hurricane | Alabama | Florida | Louisiana | Mississippi | Texas | Total |
|---|---|---|---|---|---|---|
| Katrina, Rita, and Wilma | $1,022 | $233 | $31,016 | $10,058 | $1,900 | $44,229 |
| Ike | $15 | - | $329 | - | $4,178 | $4,522 |
| Gustav | $19 | $8 | $1,544 | $47 | - | $1,618 |
| Total | $1,056 | $241 | $32,889 | $10,105 | $6,078 | $50,369 |

Source: Federal Emergency Management Agency, *Disaster Relief Fund: Monthly Report*, February 5, 2013.

## FEMA Mission Assignments by Federal Entity

Mission assignments are directives from FEMA (on behalf of the requesting state) to other federal agencies to perform specific work in disaster operations on a reimbursable basis. The mission assignment contains information that is used by FEMA management to evaluate requests for assistance from states, other federal agencies, and internal FEMA organizations.[95] **Table 18** contains a list of mission assignments by entity for Hurricanes Katrina, Wilma, and Rita. **Table 19** contains mission assignment data for Hurricane Gustav and **Table 20** contains mission assignments for Hurricane Ike. As shown in **Tables 18**, **19**, and **20**, mission assignment can be assigned directly to an agency, directly to an agency's program/activity, or both.

### Table 18. Mission Assignments by Agency: Hurricanes Katrina, Wilma, and Rita (Net Obligations, as of January 1, 2013)

| Department/Agency | Obligations |
| --- | --- |
| **Department of Agriculture** | $2,573,496 |
| Animal and Plant Health Inspection Service | $55,776 |
| Food and Nutrition Service | $10,493 |
| U.S. Forest Service | $162,523,398 |
| **Department of Commerce** | $2,171,004 |
| National Oceanic and Atmospheric Administration | $2,503,387 |
| **Department of Defense** | $380,614,318 |
| Army Corps of Engineers-Great Lakes and Ohio River Division | $2,522 |
| Army Corps of Engineers-Mississippi Valley Division | $3,606,709,470 |
| Army Corps of Engineers-South Atlantic Division | $234,037,021 |
| Army Corps of Engineers-South Western Division | $208,521,382 |
| National Geospatial Intelligence Agency | $1,005,796 |
| **Department of Energy** | $209,373 |
| **Department of Health and Human Services** | $74,004,453 |
| Centers for Disease Control and Prevention | $15,101,893 |
| **Department of Homeland Security** | |
| Customs and Border Protection | $15,487,544 |
| Federal Law Enforcement Training Center | $459,381 |
| Federal Protective Service | $182,228,449 |
| National Communications System | $4,310,150 |
| Transportation Security Administration | $351,511 |
| U.S. Citizenship and Immigration Services | $304,257 |
| U.S. Coast Guard | $183,542,905 |

| Department/Agency | Obligations |
| --- | --- |
| U.S. Immigration and Customs Enforcement | $7,7,487,035 |
| U.S. Secret Service | $8,800 |
| **Department of Housing and Urban Development** | $41,700,880 |
| **Department of Justice** | $29,976,879 |
| U.S. Parole Commission | $2,056,790 |
| **Department of Labor** | $925,851 |
| Occupational Safety and Health Administration | $4,958,193 |
| **Department of State** | $18,101 |
| **Department of the Interior** | $234,730 |
| Bureau of Indian Affairs | $21,189 |
| Bureau of Reclamation | $820,442 |
| National Park Service | $52,921 |
| U.S. Geological Survey | $471,065 |
| **Department of Transportation** | $442,007,004 |
| Federal Aviation Administration | $7,433 |
| **Department of the Treasury** | $1,754,433 |
| **Department of Veterans Affairs** | $2,931,612 |
| **Agency for International Development** | $1,749,789 |
| **American Red Cross** | $11,159 |
| **Corporation for National and Community Service** | $1,028,304 |
| **Environmental Protection Agency** | $264,062,645 |
| **Equal Employment Opportunity Commission** | $354,546 |
| **General Services Administration** | $56,410,169 |
| **National Aeronautics and Space Administration** | $1,768,211 |
| **National Archives and Records Administration** | $434,350 |
| **National Capital Planning Commission** | $7,469 |
| **National Labor Relations Board** | $215,543 |
| **Office of Personnel Management** | $400,000 |
| **Office of the Chief Financial Officer** | $70,100 |
| **Railroad Retirement Board** | $5,419 |
| **Social Security Administration** | $817,509 |
| **Tennessee Valley Authority** | $9,039,858 |
| **U.S. Postal Service** | $129,208 |
| **Total** | $5,941,178,581 |

Source: Unpublished data provided by FEMA. Data available upon request.

Notes: Mission Assignments were given to departments as well as the entities within some of the departments. The obligations data in the table reflect both department-wide and sub-department entity-specific obligations for mission assignments. Totals are not provided for each agency.

### Table 19. Mission Assignments by Agency: Hurricane Gustav (Net Obligations, as of January 1, 2013)

| Department/Agency | Obligations |
| --- | --- |
| **Department of Agriculture** | $45,000 |
| Animal and Plant Health Inspection Service | $178,465 |
| U.S. Forest Service | $2,750,000 |
| **Department of Commerce** | |
| National Oceanic and Atmospheric Administration | $15,000 |
| **Department of Defense** | |
| Army Corps of Engineers-Mississippi Valley Division | $105,349,225 |
| Army Corps of Engineers-South Atlantic Division | $1,587,780 |
| Army Corps of Engineers-South Western Division | $831,710 |
| National Geospatial-Intelligence Agency | $62,000 |
| **Department of Energy** | $120,000 |
| **Department of Health and Human Services** | $17,476,000 |
| **Department of Homeland Security** | |
| Customs and Border Protection | $857,000 |
| Federal Communications Commission | $75,000 |
| Federal Protective Service | $7,653,644 |
| Information Analysis and Infrastructure Protection Program | $220,000 |
| National Communications System | $48,426 |
| Transportation Security Administration | $13,978 |
| U.S. Coast Guard | $571,960 |
| U.S. Immigration and Customs Enforcement | $82,411 |
| **Department of Housing and Urban Development** | $140,000 |
| **Department of Justice** | $1,281,144 |
| **Department of Labor** | $10,000 |
| Occupational Safety and Health Administration | $35,000 |
| **Department of State** | $40,000 |
| **Department of the Interior** | $20,000 |
| National Park Service | $300,000 |
| **Department of Transportation** | $621,904 |
| **Department of Treasury** | $50,000 |
| **Department of Veterans Affairs** | $10,000 |
| **Corporation for National and Community Service** | $252,049 |
| **Environmental Protection Agency** | $12,007,379 |
| **General Services Administration** | $4,274,543 |
| **Tennessee Valley Authority** | $3,448,894 |
| **Total** | **$161,322,222** |

Source: Unpublished data provided by FEMA. Data available upon request.

Notes: Mission assignments were given to departments as well as the entities within some of the departments. The obligations data in the table reflect both department-wide and sub-department entity-specific obligations for mission assignments. Totals are not provided for each agency.

### Table 20. Mission Assignments by Agency: Hurricane Ike (Net Obligations, as of January 1, 2013)

| Department/Agency | Obligations |
|---|---|
| **Department of Agriculture** | $2,153,188 |
| U.S. Forest Service | $18,990,000 |
| **Department of Defense** | $25,030,836 |
| Army Corps of Engineers-Mississippi Valley Division | $19,200,000 |
| Army Corps of Engineers-South Atlantic Division | $7,926 |
| Army Corps of Engineers-South Western Division | $243,230,000 |
| National Geospatial-Intelligence Agency | $808,051 |
| **Department of Energy** | $235,000 |
| **Department of Health and Human Services** | $36,630,000 |
| **Department of Homeland Security** | |
| Customs and Border Protection | $580,000 |
| Federal Protective Service | $24,995,000 |
| Information Analysis and Infrastructure Protection Program | $340,000 |
| National Communication System | $135,000 |
| Transportation Security Administration | $639,978 |
| U.S. Coast Guard | $668,180 |
| **Department of Housing and Urban Development** | $1,346,668 |
| **Department of Justice** | $386,398 |
| **Department of Labor** | |
| Occupational Safety and Health Administration | $30,000 |
| **Department of the Interior** | $850,000 |
| Bureau of Indian Affairs | $10,000 |
| U.S. Geological Survey | $558,485 |
| **Department of Transportation** | $115,597 |
| Federal Aviation Administration | $250,000 |
| **Department of Treasury** | $4,011 |
| **Department of Veterans Affairs** | $260,000 |
| **Corporation for National and Community Service** | $84,236 |
| **Environmental Protection Agency** | $58,365,000 |
| **General Services Administration** | $1,026,351 |
| **Tennessee Valley Authority** | $4,350,768 |
| **Total** | $441,280,673 |

Source: Unpublished data provided by FEMA. Data are available upon request.

Notes: Mission assignments were given to departments as well as the entities within some of the departments. The obligations data in the table reflect both department-wide and sub-department entity-specific obligations for mission assignments. Totals are not provided for each agency.

## Community Disaster Loan Program[96]

### Community Disaster Loan Program Authority

The Community Disaster Loan (CDL) program is authorized by Section 417 of the Robert T. Stafford Disaster Relief and Emergency Assistance Act (hereinafter the Stafford Act).[97] The CDL program allows the President to:

> make loans to any local government which may suffer a substantial loss of tax and other revenues as a result of a major disaster, and has demonstrated a need for financial assistance in order to perform its governmental functions.[98]

As with most other authorities in the Stafford Act, this authority has been delegated from the President to the Administrator of FEMA.[99]

### Community Disaster Loan Program Description

Following a major disaster, the financial capacity of a local government may be severely undermined by a decrease in local revenues. The reduction in tax or other revenue can limit the local government's ability to maintain public services or afford many extraordinary but necessary expenditures in the disaster recovery process. Though there are many disaster assistance programs available to communities, the CDL program is the only program that specifically provides assistance to local governments to help compensate for revenue shortfalls. The core purpose of these community disasters loans, as detailed in the original Senate committee report authorizing the program, is "to permit the local governments to continue to provide municipal services, such as the protection of public health and safety and the operation of the public school system."[100] Consistent with this stated purpose, local governments that are eligible for loans include entities such as special districts and school boards that provide a wide array of local government services.[101]

After severe disasters, the revenue base of a local government may take years to fully recover, if ever. Some key sources of revenue may never return to pre-disaster levels, such as property taxes from areas severely damaged by a disaster. To account for the local government's continuing need for financial assistance in these circumstances, FEMA also has the authority to cancel the repayment on all or part of the loan. FEMA may cancel a loan up to the amount that the local government's tax and other revenues are insufficient to meet its cumulative operating budget over a period of three full fiscal years following a major disaster, in addition to any unreimbursed disaster-related expenses made in that time period.[102] FEMA may also forgive any amount of related interest owed on the cancelled principal of a loan.[103]

Following Hurricane Katrina and the 2005 hurricane season, the 109th and 110th Congresses appropriated funds and created unique statutory guidelines for the loan program applicable exclusively for those disasters. Collectively, these loans are referred to as "special" community disaster loans (SCDLs). FEMA promulgated regulations to govern the implementation of SCDLs that were very similar to those for the traditional loans, but with several notable distinctions.[104] These new statutory and regulatory guidelines altered some of the eligibility, size, and cancellation criteria applicable to the SCDLs. The criteria for cancelling SCDLs for Hurricane Katrina were further amended by Congress in Section 564 of P.L. 113-6, the Consolidated and Further Continuing Appropriations Act, 2013.

A similar pattern was followed by the 110th and 111th Congress for major disasters occurring in the 2008 calendar year, including Hurricanes Ike and Gustav.[105] Following an appropriation to the DADLP account in the 110th Congress (P.L. 110-329), two subsequent laws passed in the 111th Congress established several unique conditions for eligibility and loan size specific to disasters in the 2008 calendar year. FEMA has not issued new regulations in response to the unique provisions of the 2008 loans, as they were relatively minor reforms.

**Funding for the Community Disaster Loan Program**

Unlike most other Stafford Act programs, the CDL program is not funded through the DRF. The program is instead funded through the Disaster Assistance Direct Loan Program (DADLP) account.[106] The account also funds activities under Section 319 of the Stafford Act, which provides advances or loans for the portion of assistance applicants are responsible for under the different cost-sharing provisions of the Stafford Act (commonly referred to as the applicant/state "match" for assistance).[107] Generally, funds have been annually appropriated to the DALDP account for the purposes of Section 319 of the Stafford Act. However, funds for the purposes of the CDL program have often been appropriated through emergency supplemental appropriation bills in response to a particular set of disasters. Like the DRF, funds appropriated to the DALDP have traditionally been treated as "no year" funds and were available until expended. **Table 21** provides a list of the appropriations for the program relevant to Hurricanes Katrina, Rita, Wilma, Gustav, and Ike.

**Table 21. Appropriations to the DALDP account for Community Disaster Loans, 2006-2008 (Dollars in Thousands)**

| Fiscal Year | Public Law and U.S. Statutes Citation | Appropriated Amount |
|---|---|---|
| 2006 | P.L. 109-88;a 119 Stat. 2061 | $750,000 |
| 2006 | P.L. 109-234;b 120 Stat. 459-460 | $279,800 |
| 2008 | P.L. 110-329;c 122 Stat. 3592 | $98,150 |

Source: CRS analysis of enacted appropriation bills.

Notes: Appropriations to other accounts that may have funded community disaster loans through reprogramming requests not noted in statute are not captured in this table. This table does not catalogue appropriations to the DALDP for administrative expenses.

[a] The appropriation was available to governments under any major disaster declarations, but all of the appropriated money went to governments in Louisiana or Mississippi following Hurricane Katrina.

[b] This appropriation was limited to governments under major disaster declarations from Hurricane Katrina and the 2005 hurricane season, but only was used by governments in Louisiana or Mississippi following Hurricane Katrina.

[c] This appropriation was available until expended, but was used by some governments affected by Hurricanes Gustav and Ike.

FEMA's authority to cancel repayment on loans makes it difficult to calculate the true "cost" of the program, or indeed what should be considered as an obligation or expenditure from the program. There are numerous financial figures that may be of use to Congress when evaluating the program:

- the total amount appropriated to the DALDP account to subsidize loans issued by the program, and the associated figure on how much this amount supports in total loan authority;[108]
- the total principal of loans approved by FEMA for local governments to borrow;
- the total principal of loans borrowed by local governments from approved amounts;
- the total principal of loans "cancelled" for repayment by FEMA, and the inverse of this figure; and
- the total repaid by local governments and their outstanding or defaulted debt.

For brevity, this report only provides the total principal of loans borrowed by local governments following the major disasters listed. This information is displayed in **Table 22**. The true cost to the federal taxpayer from this program is difficult to evaluate, but it roughly equates to the total dollar amount (principal and interest) of loans that have been cancelled for repayment by FEMA or will not otherwise be repaid by the local governments. This figure would be, in essence, how much the money the federal government loaned to local government without it being repaid (with interest).[109] The processes for cancelling loans issued for Hurricanes Katrina, Rita, Gustav and Ike are all ongoing.[110]

**Table 22. FEMA: Community Disaster Loan Program Borrowed Loan Amounts (Cumulative loan principal borrowed by state as of June 15, 2012; Dollars in Thousands)**

| Federal Emergency Management Agency | Alabama | Florida | Louisiana | Mississippi | Texas | Total |
|---|---|---|---|---|---|---|
| Community Disaster Loan Program | - | - | - | - | - | - |
| Hurricane Katrina[a] | - | - | $837,317 | $203,424 | - | $1,040,741 |
| Hurricane Rita[a] | - | - | - | - | - | - |
| Hurricane Wilma | - | - | - | - | - | - |
| Hurricane Gustav | - | - | $9,238 | - | - | $9,238 |
| Hurricane Ike | - | - | - | - | $35,082 | $35,082 |

Source: Calculations and categorization by CRS. Raw data provided by FEMA staff.

Notes: Figures provided in this table equate to the total principal dollar amount of loans that were borrowed by local governments in each state following each major disaster declaration. These figures do not include the interest associated with that borrowed principal.

[a] Special Community Disaster Loans were officially only issued for major disaster declarations associated with Hurricane Katrina, but because Hurricanes Katrina and Rita impacted geographically similar regions, the social benefits of the loans can be seen as assisting local governments after both disasters.

# Department of Housing and Urban Development (HUD)[111]

## *Community Development Block Grants*

**Program Authority**

The Community Development Block Grant (CDBG) program was first authorized as Title I of the Housing and Community Development Act of 1974, P.L. 93-383, (42 U.S. C. §5301, et. al).

**Program Description**

Funds are allocated by formula to states, Puerto Rico, and eligible (entitlement) communities to be used to fund eligible housing, neighborhood revitalization, and economic development activities. After funds are set aside for Indian tribes and insular areas 70% of each year's annual CDBG program appropriation must be allocated to CDBG entitlement communities, including metropolitan cities with populations of 50,000 persons or more, central cities of metropolitan areas, and statutorily defined urban counties. The remaining 30% of appropriated funds are allocated to states for distribution to non-entitlement communities.

Eligible activities must meet one of three national objectives. The activity must:

- principally benefit low or moderate income persons; or
- aid in preventing or eliminating slums or blight; or
- address an imminent threat to the health or welfare of residents of an area, including disaster relief, mitigation, and long-term recovery activities.

In addition, a state or entitlement community grantee must certify that it will expend at least 70% of its CDBG allocation over a three-year period on eligible activities principally benefiting low and moderate income persons.

In addition to allowing a state or entitlement community to fund disaster-recovery efforts under the CDBG's imminent threat national objective using CDBG regular appropriation, Congress has, at its discretion, appropriated additional supplemental CDBG funds in response to presidentially declared disasters. In addition to appropriating funds for disaster recovery activities, the statute authorizing the CDBG program grants the Department of Housing and Urban Development (HUD) the authority to waive or modify program regulations, except those relating to public notice, fair housing, civil rights, labor standards, environmental review, and the program's low and moderate income targeting requirement, when CDBG funds are used to respond to presidentially declared major disasters.[112]

Funds are allocated to states and communities to cover unmet needs not covered by state and local efforts, private insurers, and standard federal disaster programs administered by the Federal Emergency Management Agency, the Small Business Administration, and the Army Corps of Engineers. As a condition of funding, grantees are required to submit, for HUDs approval, a disaster recovery plan.

**Funding**

In the aggregate, the six states identified in **Table 23** were awarded a total of $23.971 billion in CDBG disaster relief assistance to fund disaster relief activities in response to the five hurricanes identified in the table. Nearly 60% of this amount was allocated to Louisiana while Mississippi received approximately 30% of the total.

Five of the six states included in **Table 23** received a total allocation of $19.672 billion in response to the Gulf Coast hurricanes of 2005. Louisiana received the largest share (75%) of this amount followed by Mississippi (28%), Texas (2.5%), Florida (1%), and Alabama (less than 1%).

A total of $4.296 billion was awarded to five of six states included in **Table 23** to support disaster recovery activities in response to Hurricane Ike. Texas accounted for 71% of the total followed by Louisiana (25%), Tennessee (2%), Florida (1.8%), and Mississippi (less than 1%).

*Rental Assistance/Section 8 Vouchers*

The Section 8 Housing Choice Voucher program, authorized at 42 U.S.C. §1437f(o), provides portable rent subsidies that low-income families can use to rent housing units offered by private market landlords. Families with vouchers contribute an income-based payment towards their rent (generally equal to 30% of a family's income), and the federal government, through local public housing authorities (PHAs), pays the landlord the difference between the tenant's contribution and the contract rent for the unit.

Congress provided over $555 million to HUD to provide rental assistance (in the form of Section 8 Housing Choice Vouchers) to families displaced by Hurricanes Katrina and Rita. The first $390 million of that amount was appropriated to HUD to provide temporary rental assistance vouchers to families that were previously assisted by HUD programs, but were displaced by the 2005 hurricanes.

**Table 23. Distribution of CDBG Disaster Recovery Funds for Selected States, by Disaster Declaration (Allocations as of Feb. 25, 2013; Dollars in Thousands)**

| Hurricane | Alabama | Florida | Louisiana | Mississippi | Tennessee | Texas | Total |
|---|---|---|---|---|---|---|---|
| Katrina-Rita-Wilma | $95,614 | $182,970 | $13,410,000 | $5,481,221 | - | $503,194 | $19,672,999 |
| Gustav | - | - | - | $2,281 | - | - | $2,281 |
| Ike | - | $81,063 | $1,058.690 | $6,283 | $92,517 | $3,057,919 | $4,296,472 |
| Total | $95,614 | $264,033 | $14,468,690 | $5,489,785 | $92,517 | $3,561,113 | $23,971,752 |

Source: HUD, available at http://portal.hud.gov/hudportal/HUD?src=/program_offices/comm_planning/communitydevelopment/programs/dr si/activegrantee

Notes: Allocations for Hurricanes Katrina, Wilma, and Rita were reported and presented as an aggregated total.

Later, HUD was given a mission assignment by FEMA to begin providing rental assistance to all remaining households displaced by the 2005 hurricanes. HUD named this program the Disaster Housing Assistance Program (DHAP), and the cost of the DHAP was covered by FEMA's Disaster Relief Fund. Following Hurricane Ike, FEMA and HUD established another Disaster Housing Assistance Program (DHAP-Ike) for families displaced by that storm, also funded through the DRF under a mission assignment.

Following the first appropriation, and establishment of the mission assignments, Congress appropriated $85 million for HUD to fund the cost of ongoing, permanent Section 8 rental assistance vouchers for displaced families whose temporary housing assistance under DHAPKatrina was expiring. Congress later appropriated an additional $80 million to create new Section 8 rental assistance vouchers in the areas affected by Hurricanes Katrina and Rita.

**Table 24** provides the total appropriations for disaster-related rental assistance vouchers. It does not provide allocations by state for all rental assistance funding because that information is not readily available and would be difficult to determine. Most of the funding for rental assistance was not allocated to the local public housing authorities (PHAs) administering the program by state. Rather, it was allocated based on where displaced families were living. For example, a PHA in Texas may have been administering a voucher on behalf of the Housing Authority of New Orleans for a family who was living in New Orleans before the storm, but relocated to Alabama after the storm.[113] The $80 million for new vouchers was allocated to housing authorities and **Table 24** provides a break-down by state for those funds.

## *Supportive Housing*

The Louisiana Recovery Corporation titled its recovery plan, which was primarily funded with emergency CDBG funding, the "Road Home" program. As shown in **Table 24**, Congress appropriated $73 million to HUD for allocation to Louisiana's Road Home program (Supportive Housing) to fund the creation of permanent supportive housing units for the elderly and persons with disabilities. Of that amount, $50 million was appropriated through an existing homeless assistance grant program that serves homeless persons with disabilities (called Shelter Plus Care) (authorized at 42 U.S.C. Chapter 119) and $23 million was appropriated through the Section 8 Housing Choice Voucher program.[114]

## Public Housing Repair

Low-rent public housing is federally subsidized housing owned and operated by local PHAs and available to low-income families. Several public housing developments, particularly in New Orleans, suffered severe damage after Hurricane Katrina. As shown in **Table 24**, Congress appropriated $15 million in emergency funding to HUD's public housing capital fund (authorized at 42 U.S.C. §1437g), which was allocated to PHAs to aid in the repair of severely damaged public housing in Louisiana.

## Inspector General

As shown in **Table 24**, Congress appropriated $7 million for the HUD Inspector General to help fund the cost of enhanced oversight over disaster recovery funding.

**Table 24. Disaster Relief Funding by the Department of Housing and Urban Development (Allocations; Dollars in Thousands)**

| Department of Housing And Urban Development | Alabama | Florida | Louisiana | Mississippi | Texas | Total |
|---|---|---|---|---|---|---|
| Rental Assistance/Section 8 Vouchers[a] | $6,109 | $10,980 | $16,908 | $16,797 | $27,706 | $78,500 |
| Supportive Housing | - | - | $73,000 | - | - | $73,000 |
| Public Housing Repair | - | - | $15,000 | - | - | $15,000 |
| Inspector General[b] | N/A | N/A | N/A | N/A | N/A | $7,000 |
| Total | $6,109 | $10,980 | $104,908 | $16,797 | $27,706 | $173,500 |

Source: Table prepared by CRS. Figures are based on P.L. 109-148, P.L. 109-234, P.L. 110-28, P.L. 110-116, P.L. 110-252, P.L. 110-329, and P.L. 111-32. Community Development Block Grant allocations taken from http://portal.hud.gov/hudportal/HUD?src=/program_offices/comm_planning/communitydevelopment/programs/drsi/activegrantee. Rental Assistance/Section 8 Voucher allocations taken from http://www.hud.gov/offices/pih/publications/sec1203/thu-cong-ntf.pdf.

Public Housing repair information was taken from HUD's FY2010 Congressional Budget Justifications.

Notes: Total amounts allocated do not equal total amounts appropriated in some cases because funds have been reserved by the department for administrative costs.

[a] Note that state allocations are only provided for the $80 million provided for new vouchers by P.L. 111-32.

[b] An additional $9 million provided by P.L. 109-234 for Community Development Block Grants was required to be transferred to the Office of Inspector General.

# Department of Justice[115]

Established by an "Act to Establish the Department of Justice"[116] with the Attorney General at its head, the Department of Justice (DOJ) provides counsel for the government in federal cases and protects citizens through law enforcement. It represents the federal government in all proceedings, civil and criminal, before the U.S. Supreme Court. In legal matters, generally, the department provides legal advice and opinions, upon request, to the President and executive branch department heads.

To date, the DOJ has received a total of $287.5 million in supplemental appropriations for departmental expenses related to hurricanes in the Gulf of Mexico and to award grants to Gulf Coast states. **Table 25** provides a breakdown of how DOJ obligated disaster funding amongst Alabama, Florida, Louisiana, Mississippi, and Texas.

## *Legal Activities*

### Program Authority or Authorities

Subtitle A of Title XI of the Violence Against Women and Department of Justice Reauthorization Act of 2005 (P.L. 109-162) authorized appropriations for the General Legal Activities and U.S. Attorneys accounts. For the General Legal Activities account the act authorized $679.7 million for FY2006, $706.8 million for FY2007, $735.1 million for FY2008, and $764.5 million for FY2009. For the U.S. Attorneys account the act authorized $1.626 billion for FY2006, $1.691 billion for FY2007, $1.795 billion for FY2008, and $1.829 billion for FY2009.

### Program Description

The Legal Activities account includes several sub-accounts, including General Legal Activities and the U.S. Attorneys. The General Legal Activities sub-account funds the Solicitor General's supervision of DOJ's conduct in proceedings before the Supreme Court. It also funds several departmental divisions (tax, criminal, civil, environment and natural resources, legal counsel, civil rights, INTERPOL, and dispute resolution). The U.S. Attorneys enforce federal laws through prosecution of criminal cases and represent the federal government in civil actions in all of the 94 federal judicial districts.[117]

## Funding Narrative

Since 2005, Congress has appropriated a total of $17.5 million in supplemental appropriations for this account. This amount included $2.0 million for General Legal Activities and a total of $15.5 million for the U.S. Attorneys. Chapter 8 of Title II of the Emergency Supplemental Appropriations Act for Defense, the Global War on Terror and Hurricane Recovery, 2006 (P.L. 109-234) provided $2 million for General Legal Activities "to investigate and prosecute fraud cases related to hurricanes in the Gulf Coast region."[118] Chapter 8 of Title I of Division B of the Department of Defense, Emergency Supplemental Appropriations to Address Hurricanes in the Gulf of Mexico, and Pandemic Influenza Act, 2006 (P.L. 109-148) provided $9 million for the U.S. Attorneys "to support operational recovery from hurricane-related damage in the Gulf Coast region."[119] Chapter 8 of Title II of the Emergency Supplemental Appropriations Act for Defense, the Global War on Terror and Hurricane Recovery, 2006 (P.L. 109-234) provided the U.S. Attorneys with $6.5 million "to investigate and prosecute fraud cases related to hurricanes in the Gulf Coast region."[120]

## *United States Marshals Service*

### Program Authority or Authorities

Subtitle A of Title XI of the Violence Against Women and Department of Justice Reauthorization Act of 2005 (P.L. 109-162) authorized $800.3 million for FY2006, $832.3 million for FY2007, $865.6 million for FY2008, and $900.2 million for FY2009 for the United States Marshals Service (USMS) account.

### Program Description

The federal marshals' service was founded in 1789, making it the oldest federal law enforcement agency.[121] A presidentially appointed U.S. marshal directs the operations of the marshals' services in each of the 94 federal judicial districts.[122] The USMS facilitates the functioning of the federal judicial process by providing protection for judges, attorneys, witnesses, and jurors and providing physical security in courthouses.[123] The USMS is the federal government's primary agency for fugitive investigations.[124] USMS task forces combine the efforts of federal, state and local law enforcement agencies to locate and arrest fugitives.[125] The Marshals Service also works with international law enforcement agencies to apprehend fugitives who have fled abroad and to apprehend foreign fugitives who have entered the United

States.[126] The USMS executes all federal arrest warrants.[127] The USMS manages and sells assets which were seized or forfeited by federal law enforcement agencies.[128] The assets managed and sold by the USMS are assets that represent the proceeds of, or were used to facilitate federal crimes.[129] The Marshals Service is responsible for housing and transporting all federal detainees from the time they are arrested until they are either acquitted or convicted and delivered to their designated federal prison.[130] The USMS operates the Justice Prisoner and Alien Transportation System (JPATS), which transports prisoners between judicial districts, correctional facilities, and foreign countries.[131] The USMS is also responsible for administering the federal witness security program, which provides for the security and safety of government witnesses and their authorized family members, whose lives are in danger as a result of their cooperation with the U.S. government.[132]

**Funding Narrative**

Since 2005, Congress has appropriated $9 million in supplemental appropriations for the U.S. Marshal's Service. Chapter 8 of Title I of Division B of the Department of Defense, Emergency Supplemental Appropriations to Address Hurricanes in the Gulf of Mexico, and Pandemic Influenza Act, 2006 (P.L. 109-148) provided $9 million for the USMS's salaries and expenses account "to support operational recovery from hurricane-related damage in the Gulf Coast region."[133]

*Federal Bureau of Investigation*

**Program Authority or Authorities**

Subtitle A of Title XI of the Violence Against Women and Department of Justice Reauthorization Act of 2005 (P.L. 109-162) authorized $5.761 billion for FY2006, $5.992 billion for FY2007, $6.231 billion for FY2008, and $6.481 billion for FY2009 for the Federal Bureau of Investigation (FBI) account.

**Program Description**

The FBI was founded in 1908.[134] Its headquarters is in Washington, DC, and it has 56 field offices located in major cities throughout the United States and another 380 resident agencies in cities and towns across the nation.[135] In addition, the FBI has more than 60 international offices called "legal attachés" in U.S. embassies worldwide.[136] The FBI is the lead federal investigative agency charged with defending the country against foreign terrorist and intelligence threats; enforcing federal criminal laws; and providing leadership and criminal justice services to federal, state, municipal, tribal, and territorial law enforcement agencies and partners.[137] The FBI focuses on protecting the United States from internal and external threats and investigations that are too large or too complex for state and local authorities to handle on their own.[138] The priorities of the FBI include:

- protecting the United States from terrorist attack;
- protecting the United States against foreign intelligence operations and espionage;
- protecting the United States against cyber-based attacks and high-technology crimes;
- combating public corruption;
- protecting civil rights;
- investigating transnational/national criminal organizations and enterprises;
- investigating major white-collar crime;
- investigating significant violent crime; and
- supporting federal, state, local and international partners.[139]

The FBI collects and disseminates national crime data through the Uniform Crime Reports (UCR).[140] The FBI also operates several national law enforcement information sharing systems such as the Combined DNA Index System (CODIS),[141] the Law Enforcement National Data Exchange (N-Dex),[142] the Integrated Automated Fingerprint Identification System (IAFIS),[143] the National Instant Criminal Background Check System (NICS),[144] and the National Crime Information Center (NCIC).[145]

**Funding Narrative**

Since 2005, Congress has appropriated $45 million in supplemental appropriations for the FBI. Chapter 8 of Title I of Division B of the Department of Defense, Emergency Supplemental Appropriations to Address Hurricanes in the Gulf of Mexico, and Pandemic Influenza Act, 2006 (P.L. 109-148) provided $45 million for the FBI's salaries and expenses account for "to support operational recovery from hurricane-related damage in the Gulf Coast region."[146]

## Drug Enforcement Administration

**Program Authority or Authorities**

Subtitle A of Title XI of the Violence Against Women and Department of Justice Reauthorization Act of 2005 (P.L. 109-162) authorized $1.716 billion for FY2006, $1.785 billion for FY2007, $1.856 billion for FY2008, and $1.930 billion for FY2009 for the Drug Enforcement Administration (DEA) account.

**Program Description**

The DEA was established in 1973 through an executive order issued by President Nixon.[147] The DEA has 226 domestic and 85 foreign offices.[148] The DEA's mission is "to enforce the controlled substances laws and regulations of the United States and bring to the criminal and civil justice system of the United States, or any other competent jurisdiction, those organizations and principal members of organizations, involved in the growing, manufacture, or distribution of controlled substances appearing in or destined for illicit traffic in the United States; and to recommend and support non-enforcement programs aimed at reducing the availability of illicit controlled substances on the domestic and international markets."[149] The DEA's primary responsibilities include:

- investigating major violators of controlled substance laws operating at interstate and international levels;
- management of a national drug intelligence program in cooperation with federal, state, local, and foreign officials to collect, analyze, and disseminate strategic and operational drug intelligence information;
- seizure and forfeiture of assets derived from, traceable to, or intended to be used for illicit drug trafficking;
- enforcement of the provisions of the Controlled Substances Act as they pertain to the manufacture, distribution, and dispensing of legally produced controlled substances;
- reduction of illicit drugs on the United States market through methods such as crop eradication, crop substitution, and training of foreign officials; and
- liaison with the United Nations, Interpol, and other organizations on matters relating to international drug control programs.[150]

## Funding Narrative

Since 2005, Congress has appropriated $10 million in supplemental appropriations for this account. Chapter 8 of Title I of Division B of the Department of Defense, Emergency Supplemental Appropriations to Address Hurricanes in the Gulf of Mexico, and Pandemic Influenza Act, 2006 (P.L. 109-148) provided $10 million for the DEA's salaries and expenses account "to support operational recovery from hurricane-related damage in the Gulf Coast region."[151]

## *Bureau of Alcohol, Tobacco, Firearms, and Explosives*

### Program Authority or Authorities

Subtitle A of Title XI of the Violence Against Women and Department of Justice Reauthorization Act of 2005 (P.L. 109-162) authorized $923.6 million for FY2006, $960.6 million for FY2007, $999.0 million for FY2008, and $1.039 billion for FY2009 for the Bureau of Alcohol, Tobacco, Firearms, and Explosives (ATF) account.

### Program Description

The ATF enforces federal criminal law related to the manufacture, importation, and distribution of alcohol, tobacco, firearms, and explosives.[152] The ATF's responsibilities were transferred from the Department of the Treasury to the Department of Justice as a part of the Homeland Security Act (P.L. 107-296).[153] The ATF works both independently and through partnerships with industry groups, international, state, and local governments, and other federal agencies to investigate and reduce crime involving firearms and explosives, acts of arson and bombings, and illegal trafficking of alcohol and tobacco products.[154]

### Funding Narrative

Since 2005, Congress has appropriated $20 million in supplemental appropriations for the ATF. Chapter 8 of Title I of Division B of the Department of Defense, Emergency Supplemental Appropriations to Address Hurricanes in the Gulf of Mexico, and Pandemic Influenza Act, 2006 (P.L. 109-148) provided $20 million for the ATF's salaries and expenses account "to support operational recovery from hurricane-related damage in the Gulf Coast region."[155]

## Federal Prison System (Bureau of Prisons)

### Program Authority

Subtitle A of Title XI of the Violence Against Women and Department of Justice Reauthorization Act of 2005 (P.L. 109-162) authorized $5.066 billion for FY2006, $5.268 billion for FY2007, $5.479 billion for FY2008, and $5.698 billion for FY2009 for the Federal Prison System account.

### Program Description

The Bureau of Prisons (BOP) was established in 1930 to house federal inmates, to professionalize the prison service, and to ensure consistent and centralized administration of the federal prison system.[156] The BOP's mission is to protect society by confining offenders in prisons and community-based facilities that are safe, humane, cost-efficient, and appropriately secure, and that provide work and other self-improvement opportunities for inmates so that they can become productive citizens after they are released.[157] BOP currently operates 118 correctional facilities across the country.[158]

### Funding Narrative

Since 2005, Congress has appropriated $11 million in supplemental appropriations for the BOP. Chapter 8 of Title I of Division B of the Department of Defense, Emergency Supplemental Appropriations to Address Hurricanes in the Gulf of Mexico, and Pandemic Influenza Act, 2006 (P.L. 109-148) provided $11 million for the BOP's buildings and facilities account "to repair hurricane-related damage in the Gulf Coast region."[159]

## Office of Justice Programs

### Program Authorities

Congress has not traditionally authorized appropriations for the Office of Justice Programs (OJP); rather it has authorized appropriations for grant programs administered by the OJP. The funding appropriated by Congress for the OJP under the Department of Defense, Emergency Supplemental Appropriations to Address Hurricanes in the Gulf of Mexico, and Pandemic Influenza Act, 2006 (P.L. 109-148) was not appropriated pursuant to any authorized grant program. Congress appropriated funding for OJP's State and Local Law Enforcement assistance account for the OJP to award to states affected by hurricanes in the Gulf of Mexico in 2005. The funding appropriated by Congress for the OJP under the U.S. Troop Readiness,

Veterans' Care, Katrina Recovery, and Iraq Accountability Appropriations Act, 2007 (P.L. 110-28) was appropriated pursuant to an authorization for the Byrne Discretionary Grant program. This program was previously authorized under Part B of Subchapter V of Chapter 46 of Title 42 of the U.S. Code. However, the authorization was repealed by Section 1111(b)(1) of the Violence Against Women and Department of Justice Reauthorization Act of 2005 (P.L. 109-162). Congress continued to appropriation funding for the Byrne Discretionary Grant program until FY2011 when the program's funding was eliminated due to the earmark ban put in place by the 112th Congress.

**Program Description**

The OJP manages and coordinates the National Institute of Justice (NIJ), Bureau of Justice Statistics (BJS), Office of Juvenile Justice and Delinquency Prevention (OJJDP), Office of Victims of Crime (OVC), Bureau of Justice Assistance (BJA), and related grant programs. Through its component offices and bureaus, OJP disseminates knowledge and practices across America and provides grants for the implementation of crime fighting strategies.[160] NIJ focuses on research, development, and evaluation of crime control and justice issues.[161] NIJ funds research, development, and technology assistance, as well as assesses programs, policies, and technologies.[162] BJS collects, analyzes, publishes, and disseminates information on crime, criminal offenders, crime victims, and criminal justice operations.[163] BJS also provides financial and technical support to state, local, and tribal governments to improve their statistical capabilities and the quality and the utility of their criminal history records.[164] OJJDP assists local community endeavors to effectively avert and react to juvenile delinquency and victimization.[165] OJJDP seeks to improve the juvenile justice system and its policies so that the public is better protected, youth and their families are better served, and offenders are held accountable.[166] OVC distributes federal funds to victim assistance programs across the country.[167] OVC offers training programs for professionals and their agencies that specialize in helping victims.[168] BJA provides leadership and assistance to local criminal justice programs that improve and reinforce the nation's criminal justice system.[169] BJA's goals are to reduce and prevent crime, violence, and drug abuse and to improve the way in which the criminal justice system functions.[170]

**Funding Narrative**

Since 2005, Congress has appropriated $175 million for OJP for grants to assist states affected by hurricanes in the Gulf of Mexico. Chapter 8 of Title I of Division B of the Department of Defense, Emergency Supplemental Appropriations to Address Hurricanes in the Gulf of Mexico, and Pandemic Influenza Act, 2006 (P.L. 109-148) included $125 million for OJP's State and Local Law Enforcement Assistance account for "necessary expenses related to the direct or indirect consequences of hurricanes in the Gulf of Mexico in calendar year 2005." Chapter 2 of Title IV of the U.S. Troop Readiness, Veterans' Care, Katrina Recovery, and Iraq Accountability Appropriations Act, 2007 (P.L. 110-28) included $50 million under OJP's State and Local Law Enforcement Assistance Account for the Byrne Discretionary Grant program. Language in the law stated that funds provided under this program were to be used for local law enforcement initiatives in the Gulf Coast region related to the aftermath of Hurricane Katrina. Congress also required OJP to award the $50 million it received under the U.S. Troop Readiness, Veterans' Care, Katrina Recovery, and Iraq Accountability Appropriations Act, 2007 based upon each affected state's level of reported violent crime in 2005.

**Table 25. Disaster Relief Funding for the Department of Justice (Obligations: Dollars in Thousands)**

| Department of Justice | Alabama | Florida | Louisiana | Mississippi | Texas | Other | Total |
|---|---|---|---|---|---|---|---|
| Criminal Division | - | - | $440 | - | - | $935 | $1,375 |
| Civil Division | - | - | - | - | - | $625 | $625 |
| U.S. Attorneys | $79 | $1,019 | $8,806 | $3,545 | $627 | $1,424 | $15,500 |
| U.S. Marshals Service | - | $105 | $3,002 | $1,066 | $1,830 | $2,995 | $9,000 |
| Federal Bureau of Investigation | $993 | $439 | $18,469 | $674 | $107 | $24,318 | $45,000 |
| Drug Enforcement Administration | $1,906 | $2 | $4,302 | $1,848 | $135 | $1,807 | $10,000 |
| Bureau of Alcohol, Tobacco, Firearms, and Explosives | - | - | - | - | - | $20,000 | $20,000 |
| Bureau of Prisons | - | $4,100 | - | - | $6,900 | - | $11,000 |
| Office of Justice Programs | $26,448 | - | $82,830 | $65,683 | $20,000 | - | $195,000 |

Source: Unpublished data provided by U.S. Department of Justice, December 12, 2012.
Notes: The "other" categories includes funds that were not allocated specifically to an individual state, but benefited recovery efforts generally; funds that were unable to be broken out by state due to incomplete financial information (ATF only); and funds that expired. Obligations for the Office of Justice Programs includes $20 million in deobligated funds from other OJP accounts. Figures have been rounded.

# Department of Labor[171]

## Workforce Investment Act (WIA) Dislocated Worker Program

**National Emergency Grants**

The Employment and Training Administration (ETA) of the Department of Labor administers "federal government job training and worker dislocation programs, federal grants to states for public employment service programs, and unemployment insurance benefits. These services are primarily provided through state and local workforce development systems."[172]

The Workforce Investment Act (WIA, P.L. 105-220), whose programs are administered primarily by ETA, is the primary federal employment and training legislation. WIA authorizes several job training programs: state formula grants for Adult, Youth, and Dislocated Worker Employment and Training Activities; Job Corps; and other national programs, including the Native American Program, the Migrant and Seasonal Farmworker Program, the Veterans' Workforce Investment Program, and a series of competitive grant programs authorized under Section 171 of WIA.

ETA provides funding assistance for disaster relief activities primarily through the Dislocated Worker program, specifically by National Emergency Grants (NEG). NEGs are authorized under WIA Section 173(a) and are for employment and training assistance to workers affected by major economic dislocations, such as plant closures, mass layoffs, or natural disasters.[173] A majority of WIA funding for the Dislocated Worker program is allocated by formula grants to states (which in turn allocate funds to local entities) to provide training and related services to individuals who have lost their jobs and are unlikely to return to those jobs or similar jobs in the same industry. The remainder of the appropriation is reserved by DOL for a National Reserve account, which in part provides for the NEGs.[174]

**Funding Narrative**

The Department of Defense, Emergency Supplemental Appropriations to Address Hurricanes in the Gulf of Mexico, and Pandemic Influenza Act, 2006 (P.L. 109-148) provided $125 million in appropriations to ETA to award NEGs related to the consequences of hurricanes in the Gulf of Mexico in calendar year 2005. P.L. 109-148 specified that the appropriations were to remain available until June 30, 2006, and that the funds could be used to replace NEG funds previously obligated to the hurricane-impacted areas. In calendar year (CY) 2006, Alabama received $667 million, Louisiana $36.4

million, Mississippi $46.7 million, and Texas $64.9 million in NEG funding. The total of $148.6 million in NEG funding awarded to the five states, shown in **Table 26**, exceeds the $125 million appropriated in P.L. 109-148. In providing the award amounts and projects, ETA does not distinguish awards by funding source. Thus, some of the funding shown in **Table 26** is from the NEG funding in the regular annual WIA National Reserve appropriations.[175]

The "Emergency Supplemental Appropriations Act for Defense, the Global War on Terror, and Hurricane Recovery, 2006" (P.L. 109-234) provided $16 million in appropriations to ETA for "necessary expenses related to the consequences of Hurricane Katrina and other hurricanes of the 2005 season, for the construction, rehabilitation, and acquisition of Job Corps centers." P.L. 109-234 specified that the funds were to remain available until expended. Job Corps, which is administered by the DOL Employment and Training Administration, is primarily a residential job training program first established in 1964 that provides educational and career services to low-income individuals ages 16 to 24, primarily through contracts administered by DOL with corporations and nonprofit organizations. Most participants in the Job Corps program work toward attaining a high school diploma or a General Educational Development (GED) certificate, with a subset also receiving career technical training. Currently, there are 125 Job Corps centers in 49 states, the District of Columbia, and Puerto Rico.[176] The $16 million provided in P.L. 109-234 for construction, rehabilitation, and acquisition of Job Corps centers was most likely used for repair of the Gulfport (Mississippi) and New Orleans Job Corps centers, which were damaged during Hurricane Katrina.[177]

**Table 26. Disaster Relief Funding by the Department of Labor (Cumulative obligations as of December 2006; Dollars in Millions)**

| Department of Labor | Alabama | Florida | Louisiana | Mississippi | Texas | Total |
|---|---|---|---|---|---|---|
| Employment & Training Administration | | | | | | |
| National Emergency Grants | $667 | - | $36,400 | $46,692 | $64,859 | $148,618 |
| Office of the Secretary Job Corps | - | - | - | - | - | $16,000 |
| Total | $667 | - | $36,400 | $46,692 | $64,859 | $164,618 |

Source: CRS compilation of data from the Department of Labor Office of National Response, available at http://www.doleta.gov/neg/cy_awards_LastSix.cfm.

Notes: National Emergency Grant awards in were identified by reviewing the "project name" field of the Department of Labor Office of National Response data. Projects that identified Hurricanes Katrina, Rita, or Wilma were included. As noted in the text, only NEG awards for CY 2006 were included in this table.

## Department of Transportation[178]

DOT is the lead support agency under Emergency Support Function #1: Transportation, under the NRF. DOT reports on damage to transportation infrastructure and coordinates alternative transportation services and the restoration and recovery of the transportation infrastructure. At the time that Hurricanes Katrina, Rita, and Wilma struck, DOT also worked with FEMA in providing and coordinating transportation support, such as evacuation aid and shipping of critical supplies to the disaster area. However, by the time Gustav and Ike struck, DOT had turned over its role in evacuation aid and the shipping of critical supplies to FEMA.

During the hurricane response, DOT had only one permanent disaster program, the Federal Highway Administration Emergency Relief Program (ER). Other operating administrations, such as the Federal Aviation Administration and the Federal Transit Administration, also provided disaster assistance.

From a budgetary perspective, however, the DOT response to the Gulf Coast hurricanes may be viewed as either DOT funding or as FEMA funding provided to DOT for the mission assignment activities assumed by its operating administrations (see **Table 18**, **Table 19**, and **Table 20**). Funding by the FHWA, FAA, and FTA is briefly described below, and the cumulative total allocations to the Gulf of Mexico states are provided in **Table 27**.

### *Federal Highway Administration: Emergency Relief Program (ER)*

**ER Program Authorities**

The Federal Highway Administration's Emergency Relief Program (ER) is authorized by Title 23, U.S.C. §125 (Section 120 (e) for federal share payable).[179]

**Program Description[180]**

The ER program provides funds for the repair and reconstruction of roads on the federal-aid highway system that have suffered serious damage as a result of either (1) a natural disaster over a wide area, such as a flood, hurricane, tidal wave, earthquake, tornado, severe storm, or landslide; or (2) a catastrophic failure from any external cause (for example, the collapse of a bridge that is struck by a barge). Historically, however, the vast majority of ER funds have gone for natural disaster repair and reconstruction.

## ER Funding for Gulf Coast Hurricane Response

ER funding allocations for Hurricanes Katrina, Rita, Wilma, Gustav, and Ike totaled almost $3.2 billion. Of this amount, just over $2.8 billion has been obligated, see **Table 27**. Funding provided for hurricane relief includes funds from the program's annual $100 million authorization and from additional sums provided in supplemental or other appropriations acts. ER funds can only be used for roads and bridges on the federal-aid highway system. Repair and reconstruction costs for other damaged roads (mostly local roads and neighborhood streets) may be reimbursed by FEMA.

**Table 27. Emergency Relief Obligations for Gulf Coast Hurricanes (Obligations as of December 2012; Dollars in Thousands)**

| Hurricane | Alabama | Florida | Louisiana | Mississippi | Tennessee | Texas | Total |
|---|---|---|---|---|---|---|---|
| Katrina | $27,693 | $29,448 | $1,193,896 | $1,085,905 | - | - | $2,336,942 |
| Rita | - | $793 | - | - | - | $37,508 | $38,301 |
| Wilma | - | $271,462 | - | - | - | - | $271,462 |
| Gustav | - | - | $76,976 | $4,825 | - | - | $81,801 |
| Ike | - | - | $17,429 | - | - | $99,923 | $117,352 |
| Total | $27,693 | $301,703 | $1,288,301 | $1,090,730 | - | $137,431 | $2,845,858 |

Source: Federal Highway Administration.
Notes: Funds are obligated through a binding agreement, such as a project agreement, entered into by the Federal Highway Administration and a state.

### *Federal Aviation Administration (FAA)*

FAA has approved $110.5 million for repair and improvements to hurricane-damaged airport and air traffic control infrastructure.[181] Of this amount, $40.6 million was appropriated under the Department of Defense, Emergency Supplemental Appropriations to Address Hurricanes in the Gulf of Mexico, and Pandemic Influenza Act of 2006 (P.L. 109-148). FAA also provided Airport Improvement Program discretionary funds for airport repairs in the Gulf of Mexico states.[182]

### *Federal Transit Administration (FTA)*

The U.S. Troop Readiness Veterans' Care Katrina Recovery and Iraq Accountability Appropriations Act of 2007 (P.L. 110-28) appropriated $35 million for transit relief to the Gulf Coast states. The distribution of this funding across the Gulf Coast states is shown in **Table 28**. It is not unusual for FTA to be tasked by FEMA under a mission assignment to provide transit assistance to disaster victims. **Table 28** does not include these FEMA-reimbursed costs.

## Table 28. Disaster Relief Funding by Modal Administration/Program
### (Allocated Amounts; Dollars in Thousands)

| Department of Transportation | Alabama | Florida | Louisiana | Mississippi | Texas | Total |
|---|---|---|---|---|---|---|
| Federal Aviation Administration | $1,688 | $6,356 | $21,927 | $73,271 | $7,256 | $110,498 |
| Federal Highway Administration: Emergency Relief Program | $27,378 | $523,175 | $1,410,826 | $1,079,712 | $142,926 | $3,184,017 |
| Federal Transit Administration | $646 | $475 | $20,453 | $12,705 | $721 | $35,000 |
| Total | $29,712 | $530,006 | $1,453,206 | $1,165,688 | $150,903 | $3,329,515 |

Source: FAA Office of Government and Industry Affairs, FTA Office of Budget, FHWA.

Notes: The FAA total includes $1 million in Airport Improvement Program funding provided for damage caused by Hurricane Ike. Totals for FAA and FTA are based on information provided to CRS as of July 13, 2010. FHWA allocations are as of January 2013. As of January 2012, FHWA began a process of identifying unobligated ER funds and withdrawing those funds no longer needed for the events for which they were allocated. Consequently, these figures could change.

## Department of Veterans Affairs[183]

*Medical Center in New Orleans*

The Department of Veterans Affairs (VA) administers programs that provide benefits and other services to veterans and their spouses, dependents, and beneficiaries. The VA has three primary organizations to provide these benefits: the Veterans Benefits Administration (VBA), the Veterans Health Administration (VHA), and the National Cemetery Administration (NCA). The VHA provides medical care to eligible veterans and dependents. Hurricane Katrina caused extensive damage to the VA Medical Center in New Orleans.

**Funding Narrative**

P.L. 109-148 appropriated additional funds for necessary expenses due to the consequences of the hurricanes in the Gulf of Mexico in 2005. Funds were appropriated by category, including $198.3 million for medical services, and $26.9 million for general operating expenses, minor construction, and the National Cemetery Administration. P.L. 109-148 appropriated $367.5 million for major construction, of which $292.5 million was for a new facility in Biloxi, Mississippi, and $75 million was for advance planning and design work to replace the VA Medical Center in New Orleans.[184]

P.L. 109-234 appropriated $585.9 million for major construction by the VA, of which $550 million was for replacing the New Orleans Medical Center. P.L. 112-10 appropriated $310 million for FY2011, and P.L. 112-74 appropriated $60 million for FY2012, for the New Orleans Medical Center. The Administration did not request any new funding for this project for FY2013. The total estimated cost of replacing the VA Medical Center in New Orleans is $995 million.

The site decision for the new VA Medical Center in New Orleans was announced on November 25, 2008, and a groundbreaking ceremony was held on June 25, 2010. However, VA could not acquire all the land parcels necessary to construct the new medical center until late April 2011. The construction of the new facility began in May 2011.[185] The new medical center is estimated to be completed by December 30, 2014, and activation of the facility would occur in various phases after that.[186]

## Armed Forces Retirement Homes

### Gulfport Facility

The Armed Forces Retirement Home Trust Fund provides funds to operate and maintain the Armed Forces Retirement Homes (AFRH) in Washington, DC (also known as the United States Soldiers' and Airmen's Home), and in Gulfport, Mississippi (originally located in Philadelphia, PA, and known as the United States Naval Home). These two facilities provide long-term housing and medical care for approximately 1,600 needy veterans. The Gulfport campus, encompassing a 19-story living accommodation and medical facility tower, was severely damaged by Hurricane Katrina, and closed at the end of August 2005.

### Funding Narrative

P.L. 109-148 appropriated $65.8 million for the AFRH for expenses necessary because of the Gulf of Mexico hurricanes. Of the $65.8 million, $45 million was for advance planning and design work to replace the Gulfport, Mississippi, facility, which was nearly destroyed by Hurricane Katrina. The facility had almost 600 residents, the majority of whom were transferred to the Washington, DC, facility after the storm. P.L. 109-234 appropriated $176 million for construction of the new Gulfport facility, and consolidated an additional $64.7 million in previously appropriated funds for construction of the new facility. P.L. 110-329 and P.L. 111-117 provided additional funds ($8.0 million and $72.0 million, respectively) for construction and renovation at the Washington, DC, and Gulfport facilities (a breakdown between the facilities for the funding is not available). In October 2010, the new Gulfport facility was completed to which residents returned.

# Corporation for National and Community Service[187]

The National Civilian Community Corps (NCCC), authorized under the National and Community Service Act of 1990, as amended, is a residential program for individuals age 18 through 24 that conducts projects related to, among other things, disaster preparedness and relief and recovery efforts. The $10 million in Emergency Supplemental Funds provided for NCCC in P.L. 109-234 was used to support a range of program operations in the Gulf Region, from staff and member payroll and travel to covering communications, equipment, and supply costs.[188] Funding was used in

FY2007. Approximately $1.3 million went directly to the National Service Trust, which provides educational awards to NCCC members who complete 10 months of full-time service. The remaining $8.7 million was used to support program operations; it was not used to support a specific project or service. Instead, it was combined with the program's FY2007 appropriation of $26.8 million and allowed NCCC to direct members from all of its campuses to the Gulf Region for the recovery effort. The FY2007 appropriation, combined with the $8.7 million in supplemental funds, was used, among other things, to enable 1,063 members to serve 810,000 hours on 341 relief and recovery projects in the Gulf Region.

To support this work, NCCC partnered with numerous national and local organizations, local universities and churches, as well as local and federal government, including (but are not limited to) the American Red Cross; Habitat for Humanity; City Year Louisiana; The Salvation Army; Hands On Network; Federal Emergency Management Agency; St. Bernard Parish; Tulane, Xavier, and Dillard Universities; United Way of Acadiana, Louisiana; New Orleans Recovery School District; Christian Contractors Association, Mississippi; Council on Aging, Louisiana; Alliance for Affordable Energy; Arc of Greater New Orleans; Blackbelt and Central Alabama Housing Authority; various Boys and Girls Clubs; Mississippi Commission for Volunteers; and New Orleans Recreation Department.

# Environmental Protection Agency[189]

The U.S. Environmental Protection Agency's (EPA's) primary responsibilities include the implementation of federal statutes regulating air quality, water quality, pesticides, and toxic substances; the regulation of the management and disposal of solid and hazardous wastes; and the cleanup of environmental contamination.[190] In the case of declared disasters, FEMA may call on EPA to provide assistance to state and local governments, most notably in response to releases of hazardous materials and contaminants from a major disaster or emergency.[191]

## Hurricane Emergency Response Authorities

In addition to the authorities of a Presidential declaration under the Stafford Act, three federal laws authorized the development of the regulations that are embodied in the National Oil and Hazardous Substances Pollution Contingency Plan (NCP). These regulations serve as EPA's standing authority and plan for response to oil spills and releases of hazardous substances. The first two, the Oil Pollution Act (OPA) and Section 311 of the Clean Water Act, authorize federal emergency response to oil spills into U.S. waters, onto adjoining shorelines, or that may affect natural resources under the jurisdiction of the United States.[192] The third law, the Comprehensive, Environmental Response, Compensation, and Liability Act (CERCLA, commonly referred to as Superfund) authorizes federal emergency response to releases of hazardous substances into the environment.[193] The President's response authorities under these three laws are delegated by executive order to the Environmental Protection Agency (EPA) in the inland zone and to the U.S. Coast Guard in the coastal zone.[194] Other response authorities apply to oil released under certain circumstances not covered by the NCP.[195]

EPA also has additional emergency response roles related to protecting water infrastructure under other response plans and authorities if required. EPA is the lead federal agency for the water sector under the National Infrastructure Protection Plan.[196] EPA also has statutory "emergency powers" under the Safe Drinking Water Act to issue orders and commence civil action if a contaminant likely to enter a public water supply system poses a substantial threat to public health, and state or local officials have not taken adequate action.[197]

## *EPA Hurricane Response*

EPA's primary disaster response role is carried out in accordance with the National Oil and Hazardous Substances Pollution Contingency Plan (NCP)[198] as outlined in the NRF, Emergency Support Function 10 (ESF#10)—Oil and Hazardous Materials Annex. Under ESF#10, EPA is the lead federal agency for inland incidents and those affecting both inland and coastal zones.[199] EPA also has various other response roles under the NRF and may perform a wide array of support functions in responding to a disaster or emergency.[200] In accordance with various ESFs, EPA support to other federal agencies (primarily FEMA and the Army Corps of Engineers) and state and local governments, include activities necessary to address threats to human health and the environment focusing on impacts to drinking water and wastewater treatment facilities and post-disaster cleanup. EPA also may support the Army Corps of Engineers in its mission under ESF #3—Public Works and Engineering Annex—to remove disaster debris[201] and cleanup of water infrastructure facilities, and to DOE under ESF #12—Energy Annex—in its effort to maintain continuous and reliable energy supplies. In practice, EPA support for this latter function has generally involved waiving environmental requirements applicable to motor vehicle fuel under the Clean Air Act. For example, as part of the federal response to hurricanes in 2005, EPA granted certain waivers under this statute in response to requests from state and local officials when significant disruptions in fuel production or distribution occurred in the wake of these natural disasters.[202]

EPA's activities following the 2005 and 2008 hurricanes included retrieval and disposal of orphan (oil) tanks and drums, the collection of household hazardous waste, and the collection of liquid and semi-liquid waste.[203] Additionally, EPA and Corps of Engineers staff conducted assessments, providing assistance to state and local government personnel to evaluate damages to public works. Steps involved in actually restoring service include drying out and cleaning engines; testing and repairing waterlogged electrical systems; testing for toxic chemicals that may have infiltrated pipes and plants; restoring pressure (drinking water distribution lines); activating disinfection units; restoring bacteria needed to treat wastes (wastewater plants); and cleaning, repairing, and flushing distribution and sewer lines. EPA also assisted local agencies with contaminated (non-hazardous) debris management activities.

*Funding Narrative*

Initially following the 2005 and 2008 hurricanes, EPA conducted assessments and provided assistance to state and local governments using existing programs and regular funding. After the initial period EPA was eligible for reimbursement by FEMA for costs associated with these efforts under a FEMA mission assignment. Funding for EPA's response to Hurricanes Katrina, Rita, Wilma, Gustav and Ike was primarily through the FEMA mission assignments and interagency agreements with FEMA. EPA indicated that of the $505 million received cumulatively through interagency agreements for its response to the five hurricanes, $497 million has been expended.[204]

In addition to the mission assignment from FEMA, EPA received a cumulative total of $21 million in emergency supplemental appropriations under P.L. 109-148 enacted December 30, 2005, and P.L. 109-234, enacted June 15, 2006. Under P.L. 109-148, EPA received $8 million in emergency supplemental FY2006 appropriations for the Leaking Underground Storage Tank Program (LUST) for necessary expenses to address the most immediate underground storage tank needs in areas affected by Hurricanes Katrina and Rita. P.L. 109-234 increased EPA's FY2006 appropriation by an additional $7 million for assessing underground storage tanks that may have leaked in affected areas, and made $6 million available through EPA's Environmental

Programs and Management (EPM) appropriations account for increased environmental monitoring, assessment, and analytical support to protect public health during the ongoing recovery and reconstruction efforts related to the consequences of the 2005 hurricane season.

**Table 29. Disaster Relief Supplemental Appropriations for the U.S. Environmental Protection Agency (EPA): P.L. 109-148 and P.L. 109-234 (FY2006 Appropriations as Received and Distributed to States; Dollars in Thousands)**

| Department/Agency/ Program | Alabama | Florida | Louisiana | Mississippi | Texas | Total |
|---|---|---|---|---|---|---|
| ENVIRONMENTAL PROTECTION AGENCY | | | | | | |
| Leaking Underground Storage Tanks (LUST) Trust Fund | $632 | - | $10,947 | $3,421 | - | $15,000 |
| Environmental Programs and Management (EPM) | $180 | $96 | $3,073 | $1,241 | - | $4,590 |
| Total | $812 | $96 | $14,020 | $4,662 | - | $19,590 |

Source: CRS interpretation of data provided by the U.S. EPA Office of the Chief Financial Officer (OCFO). through the Agency's Office of Congressional and Intergovernmental Relations (OCIR) August 2010.

Notes: P.L. 109-148 and P.L. 109-234 provided a cumulative total of $15 million in emergency funding within EPA's LUST Trust Fund account. The Alabama Department of Environmental Management (ADEM) has indicated completion of site work related to Katrina and has initiated a return of unliquidated obligations totaling $364,670. P.L. 109-234 provided $6.0 million in emergency funding within EPA's EPM account. The total amount shown in the table for EPA's EPM account does not reflect $1.4 million for EPA's Office of Research and Development and Office of Air and Radiation. Amounts in the table may not add due to rounding.

EPA reportedly provided the cumulative $15 million included in the two supplemental appropriations under the LUST program to Alabama, Louisiana, and Mississippi in the form of grants for assessment and containment of underground tanks (by statute not to exceed $85,000 per project). EPA reported no allocation of this funding to Florida or Texas. The per-state distribution was determined jointly by EPA and the affected states based on the site evaluation information available at the time. The Alabama Department of Environmental Management (ADEM) indicated completion of site work related to Katrina and initiated a return of unliquidated obligations totaling $364,670. The majority of the $6 million emergency appropriations provided within the EPA Environmental Programs and Management appropriations account was used to fund contractors for analytical and other disaster support and to purchase equipment, including replacement of expended or damaged air monitors, within Louisiana and Mississippi. Funding was also provided for similar purposes in Alabama and Florida. No EPM funding allocation was reported for Texas. EPA provided $1.4 million of the EPM supplemental funding to its Office of Research and Development and Office of Air and Radiation for continued disaster and emergency response support, including analysis in its laboratories and air monitoring, across states affected by Hurricanes Katrina and Rita.

## *EPA Regular Appropriations*

General appropriation funds available to states in the form of grants from EPA may also have been used in the 2005 and 2008 hurricane recovery efforts, in particular, capitalization grants from the Clean Water and the Drinking Water State Revolving Funds (SRFs). The SRFs are funded within the EPA's State and Tribal Assistance Grants (STAG) appropriations account. SRF grant funding is used for local wastewater and drinking water infrastructure projects, such as construction of and modifications to municipal sewage treatment plants and drinking water treatment plants, to facilitate compliance with Clean Water Act and Safe Drinking Water Act requirements, respectively. Although, following a Presidentially-declared emergency, public drinking water and wastewater utilities are eligible for FEMA supplemental federal disaster grant assistance for the repair, replacement, or restoration of disaster damaged facilities,[205] the portions of the annual fiscal year SRF grant allotments to states may have also been used to supplement these projects.

EPA allocates annual appropriations for these capitalization grants among the states based on an established formula authorized in the Clean Water Act and based on needs surveys under the Safe Drinking Water Act.[206] States must provide 20% matching funds in order to receive the federal funds. States combine their matching funds with the federal monies to capitalize their SRFs, which they use to issue low-interest or no interest loans to finance local water infrastructure projects in communities. The recipients generally must repay the loan to the issuing state. For FY2006-FY2011, the cumulative total allotment to the five Gulf States examined in this report from Clean Water SRF annual appropriations was $1.20 billion.[207]

The cumulative total during the six-year period for the Drinking Water SRF was $1.16 billion.[208] What portion of these funds was used to support projects for infrastructure affected by the five hurricanes is not known.

## The Federal Judiciary[209]

The mission of the federal courts is to protect the rights and liberties guaranteed under the Constitution. The courts are charged with interpreting and applying the law to resolve disputes through fair and impartial judgments, and ensuring fairness and equal justice for all citizens of the United States.[210]

According to the Budget Office of the Administrative Office of the U.S. Courts, Congress appropriated $18 million in emergency judiciary funding[211] for disaster relief in the aftermath of Hurricanes Katrina and Rita. These monies were obligated to (1) reimburse per diem for judges, court staff, and federal public defenders' staff who were on temporary duty assignment, and their dependents; (2) reimburse all judges and court staff who were on temporary duty assignment for travel purposes; (3) pay for furniture, equipment, and security in the temporary locations; and (4) replace furniture and equipment in courts affected by the hurricanes. **Table 30** presents the funding provided to Louisiana, Mississippi, Texas, and Florida, as well as the additional funding to the Fifth Circuit.

**Table 30. Disaster Relief Funding by the Federal Judiciary (Obligations current as of November 2012; Dollars in Thousands)**

|  | Louisiana | Mississippi | Texas | Florida | Fifth Circuit | Tota |
|---|---|---|---|---|---|---|
| The Federal Judiciary | $4,712 | $881 | $170 | $345 | $11,891 | $17,999 |

Source: Unpublished data from the Administrative Office Budget Division, available upon request.

Notes: All figures have been rounded. The Fifth Circuit encompasses the District of the Canal Zone, Louisiana, Mississippi, and Texas. The table excludes $1,360 (nominal dollars) provided to the Fourth Circuit. The Fourth Circuit encompasses Maryland, North Carolina, South Carolina, Virginia, and West Virginia.

# Small Business Administration[212]

*Disaster Assistance Program*

**Authority**

The Small Business Administration's (SBA) Disaster Assistance Program is authorized by the Small Business Act (P.L. 85-536, Section 7(b) 72 Stat. 387, as amended).

**Program Description**

The SBA's Disaster Assistance Program provides low-interest disaster loans to homeowners, renters, and businesses, as well as to private and non-profit organizations to repair or replace real estate, personal property, machinery and equipment, inventory, and business assets that have been damaged or destroyed in a declared disaster.[213]

Table 31. Small Business Administration: Number of Approved Disaster Assistance Loans For the Five Hurricanes
(Number of Total Approved Applications as of January 29, 2013)

| Small Business Administration | Alabama | Florida | Louisiana | Mississippi | Texas | Total |
|---|---|---|---|---|---|---|
| Home Loans | 2,497 | 14,021 | 86,206 | 31,243 | 15,935 | 149,902 |
| Business Loans | 360 | 2,578 | 12,921 | 4,388 | 2,545 | 22,792 |
| Economic Injury Disaster Loans | 82 | 812 | 1,801 | 335 | 410 | 3,440 |
| Total | 2,939 | 17,411 | 100,928 | 35,966 | 18,890 | 176,134 |

Source: U.S. Small Business Administration, Office of Congressional and Legislative Affairs, correspondence with CRS on January 29, 2013.

Notes: The SBA provided disaster loans to Alabama for Hurricanes Katrina, Rita, and Wilma. Alabama did not receive loans for Hurricanes Gustav and Ike. The SBA provided disaster loans to Florida for Hurricanes Katrina, Rita, and Wilma. Florida did not receive loans for Hurricanes Gustav and Ike. The SBA provided disaster loans to Mississippi for Hurricanes Katrina, Rita, Wilma, and Gustav. Mississippi did not receive loans for Hurricane Ike. The SBA provided disaster loans to Texas for Hurricanes Katrina, Rita, Wilma, and Ike. Texas did not receive loans for Hurricane Gustav.

The SBA provides three categories of loans: (1) home loans, (2) business loans, and (3) Economic Injury Disaster Loans (EIDLs). Home disaster loans help homeowners and renters repair or replace disaster-related damages to homes or personal property. SBA regulations limit home loans to $200,000 for the repair or replacement of real estate and $40,000 to repair or replace personal property. Business disaster loans help business owners repair or replace disaster-damaged property, including inventory and supplies. Business loans are limited to $2 million. EIDLs provide assistance to small businesses, small agricultural cooperatives, and certain private, nonprofit organizations that have suffered substantial economic injury resulting from a physical disaster or an agricultural production disaster. EIDLs are limited to $2 million.

**Table 31** lists the number of approved disaster loan applications by state and by type of loan for all five hurricanes. The actual number of loans made may be somewhat lower than the number of loan applications approved, because not all approved loan applications are subsequently accepted by the borrower. **Table 32** lists the amount of the approved loans, by state.

**Table 32. Small Business Administration: Approved Disaster Loan Applications by Amount
(Cumulative Loan Amounts by State as of January 29, 2013; Dollars in Thousands)**

| Small Business Administration | Alabama | Florida | Louisiana | Mississippi | Texas | Total |
|---|---|---|---|---|---|---|
| Home Loans | $96,244 | $450,170 | $5,445,887 | $2,069,160 | $686,533 | $8,747,994 |
| Business Loans | $47,052 | $412,085 | $1,526,241 | $546,417 | $324,016 | $2,855,811 |
| Economic Injury Disaster Loans | $7,221 | $48,917 | $111,486 | $19,267 | $24,277 | $211,167 |
| Total | $150,517 | $911,172 | $7,083,615 | $2,634,844 | $1,034,826 | $11,814,973 |

Source: U.S. Small Business Administration, Office of Congressional and Legislative Affairs, correspondence with CRS on January 29, 2013.

# COST-SHARES AND PROGRAMMATIC CONSIDERATIONS: HURRICANES KATRINA, WILMA, DENNIS, AND RITA[214]

## Administrative and Congressional Waivers of Cost-Shares

P.L. 110-28, the "U.S. Troops Readiness, Veterans Care, Katrina Recovery, and Iraq Accountability Appropriations Act, 2007," which provided supplemental appropriations legislation for the war in Iraq and disaster recovery from Hurricane Katrina, provided cost-share reductions for disaster assistance provided to the affected states along the Gulf Coast.[215] The reductions provided to Alabama, Florida, Louisiana, Mississippi, and Texas were among the largest ever granted.

P.L. 110-28 provided a waiver of all state and local cost-shares for four disaster assistance programs that are a part of the Stafford Act. These programs included Section 403 (Essential Assistance), Section 406 (Repair, Restoration and Replacement of Damaged Facilities), Section 407 (Debris Removal), and Section 408 (Federal Assistance to Individuals and Households). These programs are generally cost-shared in statute at 75% federal and 25% state and local with the possibility, under specified circumstances, for a 90% federal, 10% state and local ratio. Also significant was the cost-share waiver for the Other Needs Assistance Program under Section 408, which had never been waived previously. That section of Stafford states that the "Federal share shall be 75 percent."

Section 4501 of P.L. 110-28, also states in part, the following:

> (a) Notwithstanding any other provision of law, including any agreement, the Federal share of assistance, including direct Federal assistance, provided for the States of Louisiana, Mississippi, Florida, Alabama and Texas in connection with Hurricanes Katrina, Wilma, Dennis and Rita under sections 403, 406, 407, and 408 of the Robert T.
> 
> Stafford Disaster Relief and Emergency Assistance Act (42 USC 5170b, 5172, 5173, and 5174) shall be 100 percent of the eligible costs under such sections.
> 
> (b) APPLICABILITY
> 
> 1) IN GENERAL—The federal share provided by subsection (a) shall apply to disaster assistance applied for before the date of enactment of this Act.
> 
> (2) LIMITATION—In the case of disaster assistance provided under Section 403, 406 and 407 of the Robert T. Stafford Disaster Relief and Emergency Assistance Act, the Federal share provided by subsection (a) shall be limited to assistance provided for projects for which a "request for public assistance form" has been submitted.

The statutory cost-share waivers were provided for five states. Per capita damage for Louisiana, Mississippi, and Alabama from Hurricane Katrina, and for Louisiana from Hurricane Rita, had already qualified those states for a decreased state cost-share (from 25% to 10%) through FEMA's regulatory formula based on estimated damage. Congress' inclusion of Florida and Texas may have been an effort to not separate out related damages within a devastating hurricane season.

Also, the decision to grant cost-share waivers to Florida and Texas may have been in recognition of the amount of help both states had provided to Mississippi and Louisiana, respectively, in both the provision of emergency management resources and in hosting large numbers of evacuees in the wake of the storms of 2005.

The "Limitation" in the legislation was intended to ensure that the projects receiving the waiver were ones already identified by applicants and not newly created projects, or perhaps, projects not necessarily related to the event that were attempting to capitalize on the reduced cost-share provision. The legislation states that a "request for public assistance" submitted prior to enactment of the bill (May 25, 2007) will require no cost-share. Any "requests for public assistance" not submitted prior to the enactment of the bill will be cost-shared at the 90% federal, 10% state and local cost-share for the affected states. This provision appeared to be intended to provide the more generous cost-share to those projects already selected by the state rather than projects that could be developed or submitted based on 100% federal funding.

There have been several instances when Congress chose to adjust a state's cost-share by legislation. Prior to large cost-share adjustments made to several FEMA programs as noted above, Congress also legislatively reduced cost-shares for states affected by Hurricane Rita.[216]

# CONCLUDING OBSERVATIONS AND POLICY QUESTIONS[217]

This report demonstrates not only the significant amount of assistance the federal government provides for major disasters, but also the wide-range of federal programs that are brought to bear to help individuals and communities respond and recover from major disasters, as well as prepare and mitigate against future disasters. Yet, this is only a partial picture of the amounts and types of disaster assistance that have been provided by the federal government. The research focus for this report was on supplemental appropriations. The federal government also routinely provides disaster assistance through its annual appropriations. For example, from FY2004 to FY2012 Congress appropriated nearly $15 billion for the DRF (see **Table 33**). While some of the appropriated funds were used to address the damages caused by the 2005 and 2008 hurricanes, it is unclear how much was used for such purposes. In addition, many federal entities do not have a designated account for disaster relief and provide assistance that is not reimbursed by the DRF. Such entities provide disaster assistance from their regular budgets. Consequently, there is a lack of knowledge regarding the true amount of federal disaster assistance provided.

**Table 33. Disaster Relief Fund Annual Appropriations FY2004-FY2012 (Dollars in Millions)**

| Fiscal Year | Amount |
|---|---|
| 2004 | $1,800 |
| 2005 | $2,042 |
| 2006 | $1,770 |
| 2007 | $1,500 |
| 2008 | $1,400 |
| 2009 | $1,278 |
| 2010 | $1,600 |
| 2011 | $2,645 |
| 2012 | $700 |
| Total | $14,735 |

Source: CRS analysis of various Administration budget documents and appropriations statutes.

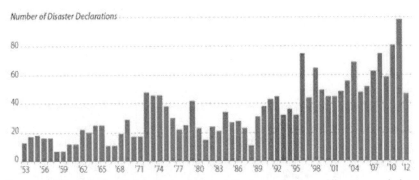

Source: CRS analysis of data derived from http://www.fema.gov/disasters and data provided by FEMA.

Figure 2. Major Disaster Declarations; 1953-2012.

Furthermore, in recent years the federal government has responded to gubernatorial requests for assistance on an increasing basis. As shown in **Figure 2**, since the use of disaster declarations began being used as a trigger for federal assistance in 1953, the number of declarations issued each year has increased significantly. The average number of major disaster declarations issued per year in the 1960s (the first full decade for declarations) was roughly

19. In contrast, from 2000 to 2009 the average number of declarations issued per year was 56. Calendar year 2011 was the busiest year on record with 99 major disaster declarations.[218]

Congressional oversight and debates concerning disaster relief can be better informed with more accurate data and information on the amounts and types of assistance provided by the federal government to the states. Thus while this report provides the most detailed information on Gulf Coast assistance, there is a need for further research on the subject of federal disaster assistance— including the assistance provided in response to disasters in other regions of the United States—to address existing gaps in funding information.

Potential policy methods for addressing gaps in funding information may include requiring:

- the issuance of disaster assistance reports on an annual or quarterly basis from all federal entities that provide significant amounts of disaster assistance;
- the Office of Management and Budget (OMB) to compile a report on an annual or quarterly basis with funding information that details all federal spending for emergencies and major disasters;
- a standardization of how expenditure data are reported across federal agencies to facilitate cost comparisons;[219]
- reports to include state-specific as well as disaster-specific information. State-specific information could be used to target mitigation projects;
- disaster assistance reports to include supplemental as well as regular appropriations data;
- federal agencies to flag monies used for disaster relief that has been taken from their regular budgets; and
- disaster assistance reports to contain cost share information as well as detailed information on state expenditures.

## Potential Methods for Controlling Costs Associated with Major Disasters

If the increase in the number of declarations and their associated costs are of concern, in addition to requiring improved data reporting Congress may choose to address the issue through a variety of policy measures.

The following sections could be used to frame a potential debate on limiting the number of declarations being issued, limiting the assistance provided after a declaration has been declared, or both.

## *Rationale for Keeping the Disaster Assistance the Same*

To many, providing relief to disaster victims is an essential role of the government. In their view, the concern over costs is understandable given concerns over the national budget. However, they may argue that the increase in the amount of assistance provided over the past decade is justified because the occurrences of disasters are on the rise (see **Figure 2**).[220] The rise may be due to a number of factors including increases in inclement weather, population growth, and building development. Moreover, proponents of keeping the current system in place may say that providing assistance to disaster-stricken areas is both acceptable and needed to help a state and region's economy recover from a storm that it otherwise may not be able to recover from on its own.

## *Offsetting the Costs of Disasters*

The DRF is the account used to fund FEMA assistance programs. The DRF receives an annual appropriation. Because the DRF is a no-year fund, any funds left over at the end of the fiscal year are carried over to the next fiscal year. However, when the account balance becomes exhausted, Congress typically has provided additional funds through a supplemental appropriation.[221] While intended to be used only occasionally, supplemental funding for the DRF has become common practice over the past decade. For example, from 2001 to 2012 in addition to regular appropriations, the DRF was replenished with 18 supplemental appropriations—an average of 1.9 per year. In some cases, the fund has received up to three additional supplemental appropriations (see **Table 34**).

The need for supplemental appropriations could have been due, in part, to the formula previously used to determine the Administrative request for the DRF. In the past, the request was based on a five-year rolling average of disaster costs—excluding large scale disasters of $500 million or more. This formula may have led to Administrative requests that were unrealistic considering the costs of contemporary disasters. The Budget Control Act (P.L. 112-25, hereinafter the BCA) provides a mechanism that allows for a higher regular appropriation for the DRF through an allowable adjustment to the BCA budget caps.[222] The increased budget authority for the account may decrease the need for frequent supplemental appropriations.

## Table 34. Major Disasters and Supplemental Appropriations FY2001-FY2012

| Fiscal Year | Disaster Event and Date of Major Disaster Declaration | Date Supplemental Appropriations Signed Into Law and P.L. Number |
|---|---|---|
| 2013 | Hurricane Sandy, Nov. - Dec. 2012 | Jan. 29, 2013, P.L. 113-2 |
| 2012 | Storms, Flooding, Drought, and Hurricane Irene events in 2011 | Dec. 23, 2011, P.L. 112-77 |
| 2010 | Hurricane Katrina, severe storms/flooding, wildfires oil spill various dates | July 19, 2010, P.L. 111-212 |
| 2008 | Hurricane, Midwest flooding and the 2008 hurricanes, various dates | Sept. 30, 2008, P.L. 110-329 |
| 2008 | Hurricane Katrina, and other hurricanes in the 2005 season | June 30, 2008, P.L. 110-252 |
| 2008 | Hurricane Katrina and California Wildfires, October 24, 2007 | Nov. 13, 2007, P.L. 110-116 |
| 2007 | Hurricane Katrina, Aug. 29, 2005 | May 25, 2007, P.L. 110-28 |
| 2006 | Hurricanes Katrina, Rita, Wilma, Aug. - Sept. 2005 | June 15, 2006, P.L. 109-234 |
| 2006 | Hurricanes Katrina, Rita, Wilma, Aug. - Sept. 2005 | Dec. 30, 2005, P.L. 109-148 |
| 2005 | Hurricane Katrina, Aug. 29, 2005 | Sept. 8, 2005, P.L. 109-62 |
| 2005 | Hurricane Katrina, Aug. 29, 2005 | Sept. 2, 2005, P.L. 109-61 |
| 2005 | Hurricanes Ivan, Jeanne, Sept. 1, 2004 | Oct. 13, 2004, P.L. 108-324 |
| 2004 | Hurricanes Charley, Frances, Sept. 1, 2004 | Sept. 8, 2004, P.L. 108-303 |
| 2004 | Wildfires, various dates | Aug. 8, 2004, P.L. 108-287 |
| 2004 | Hurricane Isabel Sept. 18, 2003 | Nov. 6, 2003, P.L. 108-106 |
| 2003 | Storms, various 2003 dates | Sept. 30, 2003, P.L. 108-83 |
| 2003 | Tornadoes, May 6, 2003 | Aug. 8, 2003, P.L. 108-69 |
| 2002 | Terrorist attacks, Sept. 11, 2001 | Aug. 2, 2002, P.L. 107-206 |
| 2001 | Terrorist attacks, Sept. 11, 2001 | Sept. 18, 2001, P.L. 107-38 |
| 2001 | Nisqually Earthquake | July 24, 2001, P.L. 107-20 |

Source: CRS analysis of appropriation statutes.

Some have proposed that supplemental funding should be "offset."[223] Appropriation legislation that is fully offset has no overall net cost in budget authority or outlays. Offsets can be achieved by cutting budget authority from one account and providing it to another, rescissions, or transferring budget authority from other programs. In recent years the debate over the use of offsets for disaster relief or assistance has intensified due to the growing size of the budget deficit and national debt.

As a result of recent congressional deliberations, legislative attempts have been made to offset the costs of disaster assistance. For example, Title VI of the House-reported version of H.R. 2017, the FY2012 Homeland Security Appropriations bill would have provided an additional $1 billion of additional funding to the DRF by transferring resources from the Department of Energy. The provision reads as follows:

> Sec. 601. Effective on the date of the enactment of this Act, of the unobligated balances remaining available to the Department of Energy pursuant to section 129 of the Continuing Appropriations Resolution, 2009 (division A of P.L. 110-329), $500,000,000 is rescinded and $1,000,000,000 is hereby transferred to and merged with 'Department of Homeland Security—Federal Emergency Management Agency—Disaster Relief': Provided, That the amount transferred by this section is designated as an emergency pursuant to section 3(c)(1) of H.Res. 5 (112$^{th}$ Congress).[224]

Another example is H.Amdt. 4 to Disaster Relief Appropriations Act, 2013 in the 113$^{th}$ Congress, which would have provided an offset of the $17 billion in emergency funding to address the immediate needs for victims and communities affected by Hurricane Sandy. The offset would have been achieved by an across-the-board rescission of 1.63% to all discretionary appropriations for FY2013.

Proponents of offsets argue that they provide a mechanism to control spending and offset the costs of disaster assistance. Opponents argue that offsets politicize disaster assistance because the program selected for the offset may have been selected because it is unpopular with a particular party. They may also argue that the debate over the use of offsets will unnecessarily slow the delivery of needed assistance. One potential argument against the sole reliance on offsets to limit federal spending on disaster assistance is that it fails to address the growing number of declarations issued each year (see **Figure 2**). As the number of declarations increase over time so too will their total cost. And as their total cost rises, more and more funding will be needed from other federal programs to fund offsets to subsidize disaster costs.

### *Limiting the Number of Major Disaster Declarations Being Issued*

Others may contend that too many major disasters are being declared and should be limited. The following sections review some policy mechanisms that could be employed to decrease the number of declarations that are being issued. The primary option consists of preventing what may be perceived by some to be marginal incidents from triggering federal assistance. Potential methods to achieve this include changing the definitions of a major disaster in Stafford Act, changing the per capita formula for determining whether a disaster is sufficiently large to warrant federal assistance, or the use of other indicators instead of, or in conjunction with, the per capita formula.

**Changing the Definition of Major Disaster in the Stafford Act**

Some argue that the Stafford Act has enhanced presidential declaration authority because the definition of a major disaster in Section 102(2) of Stafford Act is ill-defined.[225] Because of the expansive nature of this definition under the Stafford Act, they assert, there are not many restrictions on the types of major disasters for which the President may issue a declaration.[226] For example, some would argue that snowstorms do not warrant major disaster declarations.

**Changing the per Capita Formula**

One potential method of reducing the number of major disasters being declared is to increase the per capita amount used by FEMA to make major disaster recommendations to the President. A per capita formula based on damages caused by an incident is used by FEMA to make recommendations to the President concerning whether to issue a major disaster declaration. The current per capita amount used by FEMA to make recommendations is $1.32. This amount could be increased (for example, by 10%) to reduce the number of incidents eligible for federal assistance.

If increased, Congress might require that the per capita be adjusted annually for inflation. The DHS Inspector General issued a report in May 2012, which noted that FEMA had been using a $1 per capita damage amount since 1986 for determining during its preliminary damage assessment process if it would recommend to the President that the event was beyond the capacity of state and local governments to deal with without federal assistance. The DHS Inspector General also explained that FEMA did not begin adjusting that number for inflation until 1999. The DHS Inspector General pointed out that if the inflation adjustment had been occurring over that 13-year period, from 1986 to 1999, fully 36% fewer disasters would have qualified for a presidential declaration based on that factor.[227]

However, it is also useful to understand that the actual public announcement of factors considered for a declaration did not become public until 1999. At the behest of Congress, it was in that year that FEMA began to print the factors that were considered in regulation. Until then, all of that information had been within the "pre-decisional" part of the process in the executive branch. However, in 1999 FEMA began to identify factors considered for both Public and Individual Assistance. That is not to say FEMA was not using the per capita amount in its considerations, only that the process was not widely known or understood as it presently is. As the DHS IG notes, FEMA could have been raising that amount gradually, a process that did not begin until more than a dozen years later. On the other hand, it should also be considered that when FEMA discussed such proposals (e.g., per capita figures gradually increasing) with Congress, the result was a new Section 320 of the Stafford Act that stated:

> No geographic area shall be precluded from receiving assistance under this Act solely by virtue of an arithmetic formula or sliding scale based on income of population.

## *The Use of State Capacity Indicators*

In 2001, the Government Accountability Office (GAO) issued a report on disaster declaration criteria. The GAO report was a comprehensive review of FEMA's declaration criteria factors. GAO recommended that FEMA "develop more objective and specific criteria to assess the capabilities of state and local governments to respond to a disaster" and "consider replacing the per capita measure of state capacity with a more sensitive measure, such as a state's total taxable resources."

The state's Total Taxable Resources (TTR) was developed by the Department of the Treasury. GAO reported that TTR:

> is a better measure of state funding capacity in that it provides a more comprehensive measure of the resources that are potentially subject to state taxation. For example, TTR includes much of the business income that does not become part of the income flow to state residents, undistributed corporate profits, and rents and interest payments made by businesses to out-of-state stock owners. This more comprehensive indicator of state funding capacity is currently used to target federal aid to low-capacity states under the Substance Abuse and Mental Health Service Administration's block grant programs. In the case of FEMA's Public Assistance program, adjustments for TTR in setting the threshold for a disaster declaration would result in a more realistic estimate of a state's ability to respond to a disaster.[228]

It could be argued that the use of TTR would conflict with the prohibition against the use of arithmetic formulas established by Congress. However, just as FEMA's per capita measurement is one of several factors considered and not the "sole" determinant of a declaration, GAO stated that TTR would not violate Section 320 because TTR could also be used with other criteria such as those identified in regulations. Thus, some could contend that TTR could fill a similar role with perhaps more accuracy. It may also help reduce federal costs for disaster assistance by denying assistance to marginal incidents that could be otherwise handled by the state.

### *Expert Panels*

Some have proposed the use of an independent expert panel to review gubernatorial requests for major disaster declarations.[229] Such panels would be comprised of individuals with specialized knowledge in certain subject areas, such as disasters, economics, and public health. The panel would take into account the severity of the incident as well as other factors that might indicate how well the state could respond to and recover from the incident. The panel would then make recommendations to the President whether the circumstances of the incident were worthy of federal assistance based on their assessment.

Some might argue that the use of an expert panel would make decisions about whether to provide assistance more objective. Others might argue that the use of a panel may slow down the declaration process and impede the provision of important assets and resources. It may be argued that the panel's recommendation would infringe on the President's authority to issue a declaration. On the other hand, it could also be argued that the President would retain the authority to issue a declaration despite the panel's recommendation.

### *Emergency Loans*

Another potential method to reduce the number of declarations and the costs of federal disaster assistance would be to create incentives to dissuade states from requesting assistance. One method would be converting some, or all, federal assistance provided through emergency declarations into a loan program. For example, emergency declarations could be altered to provide up to a specified amount (for example, $5 billion dollars) in low interest recovery loans.[230] Under this arrangement a state could elect to handle the incident without federal assistance rather than having to reimburse the federal government for recovery loans.

## Changes to the Stafford Act

The following section discusses some potential changes to the Stafford Act that might limit the number of declarations being issued each year.

### Repeal of Section 320

As mentioned previously, Section 320 of the Stafford Act restricts the use of an arithmetic or sliding scale to provide federal assistance. Repealing Section 320 would allow formulas that establish certain thresholds that states would have to meet to qualify for assistance.

### Section 404

Section 404 of the Stafford Act[231] authorizes the President to contribute up to 75% of the cost of an incident toward mitigation measures that reduce the risk of future damage, loss of life, and suffering. Section 404 could be amended to make mitigation assistance contingent on state codes being in place prior to an event. For example, states that have met certain mitigation standards could remain eligible for the 75% federal cost share. States that do not meet the standards would be eligible for a smaller share, such as 50% federal cost share. The amendment may incentivize mitigation work on the behalf of the state and possibly help reduce damages to the extent that a request for assistance is not needed, or the cost of the federal share may be lessened. The amendment could be set to take effect in three years, giving states time to act, or not.

### Other Potential Amendments to the Stafford Act

Other amendments to the Stafford Act could either limit the number of declarations being issued, or the amount of assistance provided to the state by the federal government.

- The Stafford Act could be amended so that there could be no administrative adjustment of the cost-share. The cost-share could only be adjusted through congressional action. The amendment could be designed to apply immediately.
- The Stafford Act could be amended so that federal assistance would only be available for states with corollary programs (such as Public Assistance, Individual Assistance, and housing assistance). Establishing these programs at the state level may increase state capacity to handle some incidents without federal assistance. The

amendment could be designed to take effect in three years, giving states time to act, or not.
- The Stafford Act could be amended to discontinue all assistance for snow removal unless directed by Congress. The amendment could be designed to take effect in three years to give states and localities an opportunity to increase snow removal budgets, or not.

## *Reducing the Amount of Assistance Provided through Declarations*

### Adjust the State Cost Share

Most discussions regarding state cost-shares in disaster programs and projects involve ways in which the state amount may be reduced and the federal share increased.[232] Some may contend, however, that the opposite approach should be adopted and efforts should be undertaken to reduce disaster costs by shifting the costs to the state and local level. Currently, state and local governments provide 25% of disaster costs on projects and grants to families and individuals with the federal government assuming, at a minimum, 75% of all costs.[233]

There is no statutory limit on the number of people that can be helped following a disaster.[234] Similarly, when assessing damage to state and local infrastructure there is no cap on the amount of federal funds that can be expended to make the repairs or accomplish a replacement. The only limitation is that the damage must be to eligible facilities and that it is disaster-related damage. Given that open-ended commitment by the federal government, some may argue that increasing the state share of 25% to a higher percentage would be warranted given the federal government's fiscal condition. Another option would be to make the cost-share arrangement not subject to administrative adjustment.

### Disaster Loans

As mentioned previously, the assistance provided for emergency declarations could be provided through the form of loans. Similarly, some or all of the assistance provided to the state after a major disaster could be converted to low-interest or no-interest loans. For example, a state may receive the traditional 75% cost share for an incident but be required to reimburse 25% of that funding to the federal government. Loans for disaster recovery could also be incentivized. For instance, states that undertook certain pre-established preparedness mitigation measures could qualify for a larger federal share or a lower interest rate.

## Policy Questions

Congress has always debated the federal role in disaster relief. In recent years the debate has intensified in light of the federal budgetary environment. Policymakers have, or may ask, a number of questions relating to federal expenditures on disaster relief to assist and improve oversight, and to better inform deliberations on legislation designed to assist individuals and communities respond and recover from incidents. Such questions may include:

- To what degree should the federal government be involved in providing disaster assistance? Is the federal government providing enough assistance, or being overly generous in providing financial assistance to states?
- Was the funding provided for the Gulf Coast storms delivered efficiently and to its intended targets? If not, how can the process be improved without slowing the provision of necessary services and resources?
- How were funding allocations to each federal entity determined? Was the process accurate, or could it be improved in upcoming disasters?
- Are there increased instances of fraud, abuse, and waste when large sums of funding are provided for disaster relief? If so, what oversight mechanisms are in place to prevent such occurrences?
- Is there unnecessary duplication of services and/or efforts given the large number of federal entities involved in disaster relief?
- The assistance provided by the federal government to the Gulf Coast was provided, in part, by a number of supplemental appropriations. Is it better to provide funding overtime through multiple supplemental appropriations, or to provide the funding once through a single supplemental appropriation?

## APPENDIX A. SOURCES FOR FIGURE 1

***1871 Chicago Fire***
Wayne Blanchard, Ph.D., *Worst Disasters - Lives Lost (U.S.)*, Federal Emergency Management Agency, FEMA Emergency Management Higher Education Project, July 5, 2006.

*1900 Galveston Hurricane*
National Oceanic and Atmospheric Administration, *The Great Galveston Hurricane of 1900*, August 30, 2007, http://celebrating200years.noaa.gov /magazine/galv_hurricane/.

*1906 San Francisco Earthquake*
Wayne Blanchard, Ph.D., *Worst Disasters - Lives Lost (U.S.)*, Federal Emergency Management Agency, FEMA Emergency Management Higher Education Project, July 5, 2006.

*1919 Influenza Pandemic*
Wayne Blanchard, Ph.D., *Worst Disasters - Lives Lost (U.S.)*, Federal Emergency Management Agency, FEMA Emergency Management Higher Education Project, July 5, 2006.

*1929 Great Mississippi Flood*
Hydrologic Information Center, *Flood Losses: Compilation of Flood Loss Statistics*, National Oceanic and Atmospheric Administration/National Weather Service, Silver Spring, MD, February 1, 2011.

*1964 Alaska Earthquake/Tsunami*
United States Geological Survey, *$40^{th}$ Anniversary of "Good Friday" Earthquake Offers New Opportunities for Public and Building Safety Partnerships*, Reston, VA, March 26, 2004, http://www.usgs.gov /newsroom /article.asp?ID=106.

*1969 Hurricane Camille*
National Oceanic and Atmospheric Administration /National Weather Service, *Hurricane Camille 1969*, Flowood, MS, August 20, 2010, http://www.srh.noaa.gov/jan/?n= 1969 08 17 hurricane camille.
Edward N. Rappaport, Jose Fernandez-Partagas, and Jack Beven, *The Deadliest Atlantic Tropical Cyclones, 1492 - Present*, Appendix 1: Atlantic tropical cyclones causing at least 25 deaths, April 22, 1997, http://www.nhc.noaa.gov/pastdeadlya1.html.

*1974 Xenia (Easter) Tornado Outbreak*
National Oceanic and Atmospheric Administration, *Weather Service Commemorates Nation's Worst Tornado Outbreak*, March 31, 1999, http://www.publicaffairs.noaa.gov/storms/release.html.

## 1978 Love Canal
Eckardt C. Beck, *The Love Canal Tragedy*, Environmental Protection Agency, January 1979, http://www.epa.gov/history/topics/lovecanal/01.htm.

## 2008 Hurricane Ike
Robbie Berg, *Tropical Cyclone Report: Hurricane Ike*, National Hurricane Center, AL092008, May 3, 2010, p. 9, http://www.nhc.noaa.gov/pdf/TCR-AL092008_Ike_3May10.pdf.

## 1980 Mount St. Helens
Robert I. Tilling, Lyn Topinka, and Donald A. Swanson, *Economic Impact of the May 18, 1980 Eruption*, United States Geological Survey, Eruptions of Mount St. Helens: Past, Present, and Future: USGS Special Interest Publication, 1990.

## 1989 Loma Prieta Earthquake
Robert A. Page, Peter H. Stauffer, and James W. Hendley II, *Progress Toward A Safer Future Since the 1989 Loma Prieta Earthquake*, United States Geological Survey, U.S. Geological Survey Fact Sheet 151-99 Online Version 1.0, 1999, http://pubs.usgs.gov/fs/1999/fs151-99/.

## 1992 Hurricane Andrew
National Oceanic and Atmospheric Administration, *Famous Hurricanes of the 20$^{th}$ and 21$^{st}$ Century In the United States 1900 - 2004*, September 16, 2010.

## 1994 Northridge Earthquake
Unpublished data provided to CRS by FEMA.

## 1995 Chicago Heat Wave
Jim Angel, *The 1995 Heat Wave in Chicago, Illinois*, Illinois State Climatologist Office, Champaign, IL, http://www.isws.illinois.edu/atmos/statecli/General/1995Chicago.htm.

## 1989 Hurricane Hugo
National Oceanic and Atmospheric Administration, Famous Hurricanes of the 20$^{th}$ and 21$^{st}$ Century In the United States 1900 - 2004, September 16, 2010.

### 1994 Northridge Earthquake
United States Geological Survey, *Alaska and Washington Yield Largest U.S. Earthquakes ... Most Significant Earthquakes of '96 Rattle China, Indonesia*, February 13, 1997, http://www.usgs.gov/ newsroom/article_pf.asp?ID=975.

### 2001 September 11*th* Terrorist Attacks
National Commission on Terrorist Attacks Upon the United States, *9/11 Commission Report*, Notes on Chapter 9, Washington, DC, p. 552.

### 2005 Hurricane Katrina
Richard D. Knabb, Jamie R. Rhome, and Daniel P. Brown, *Tropical Cyclone Report*, National Oceanic and Atmospheric Administration/National Hurricane Center, Hurricane Katrina 23-30 August 2005, August 9, 2006, p. 11, http://www.nhc.noaa.gov/pdf/TCR-AL122005_Katrina.pdf.

### 2008 Hurricane Ike
National Oceanic and Atmospheric Administration/National Hurricane Center, *Hurricane History: Ike 2008,* http://www.nhc.noaa.gov/HAW2/english/history.shtml#ike.

## End Notes

[1] This section was coauthored by Bruce Lindsay, Analyst in American National Government, Government and Finance Division; and Jared Nagel, Information Research Specialist, Government and Finance Division.
[2] "Colorado State U. Review finds 2005 Hurricane Season 'Most Active,'" *Insurance Journal*, February 5, 2006.
[3] Ibid.
[4] National Oceanic and Atmospheric Administration, "Hurricane Katrina," available at http://www.katrina.noaa.gov.
[5] U.S. Congress, Senate Committee on Homeland Security and Governmental Affairs, *Hurricane Katrina: A Nation Still Unprepared*, 109th Cong., 2nd sess., S.Rept. 109-322 (Washington: GPO, 2006), p. 42.
[6] See National Oceanic and Atmospheric Administration, National Climate Data Center Billion-Dollar Weather/Climate Disaster website, available at http://www.ncdc.noaa.gov/billions/events.
[7] The White House, *The Federal Response to Hurricane Katrina: Lessons Learned*, February 23, 2006, p. 7, available at http://library.stmarytx.edu/acadlib/edocs/katrinawh.pdf.

[8] Kimberly A. Geaghan, *Forced to Move: An Analysis of Hurricane Katrina Movers*, U.S. Census Bureau, SEHSD Working Paper, Washington DC, June 2011, p. 1, available at http://www.census.gov/hhes/www/hlthins/publications/ HK_Movers-FINAL.pdf.

[9] Kristy Frame, Lynne Montgomery, and Christopher Newbury, *Bank Performance after Natural Disasters: a Historical Perspective*, Federal Deposit Insurance Corporation, January 16, 2006, available at http://www.fdic.gov/ bank/analytical/regional/ro20054q/na/2005_winter 01.html.

[10] This may have been the result of Texas and Louisiana officials evacuating over 3 million residents before Rita made landfall. See National Oceanic and Atmospheric Administration, "Hurricane Rita," August 21, 2012, available at http://www.ncdc.noaa.gov/special-reports/rita.html#impacts.

[11] National Oceanic and Atmospheric Administration, National Hurricane Center, "Hurricanes in History," available at http://www.nhc.noaa.gov/outreach/history.

[12] Ibid.

[13] Ibid.

[14] John L. Beven II and Todd B. Kimberlain, *Tropical Cyclone Report: Hurricane Gustav*, National Hurricane Center, AL072008, January 22, 2009, p. a, available at http://www.nhc.noaa.gov/pdf/TCR-AL072008_Gustav.pdf.

[15] National Oceanic and Atmospheric Administration, National Hurricane Center, "Hurricanes in History," available at http://www.nhc.noaa.gov/outreach/history.

[16] Ibid.

[17] This section was coauthored by Jared Brown, Analyst in Emergency Management and Homeland Security Policy, Government and Finance Division.

[18] U.S. Congress, House Committee on Transportation and Infrastructure, *A Review of the Preparedness, Response to and Recovery from Hurricane Sandy*, 112th Cong., 2nd sess., November 4, 2012.

[19] For example, the Department of Housing and Urban Development's Section 8 program also provides vouchers for disaster victims.

[20] P.L. 110-161, codified at 42 U.S.C. §5208.

[21] These include P.L. 109-61, P.L. 109-62, P.L. 109-148, P.L. 109-171, P.L. 109-234, P.L. 110-28, P.L. 110-116, P.L. 110-252, P.L. 110-329, and P.L. 111-32.

[22] For a discussion of funding terminology, see CRS Report 98-410, *Basic Federal Budgeting Terminology*, by Bill Heniff Jr.

[23] This section was authored by the following individuals in the Resources, Science, and Industry Division: Dennis A. Shields, Specialist in Agricultural Policy; Megan Stubbs, Specialist in Agricultural Conservation and Natural Resources Policy; Kelsi Bracmort, Specialist in Agricultural Conservation and Natural Resources Policy; Tadlock Cowan, Analyst in Natural Resources and Rural Development; and Randy Alison Aussenberg, Analyst in Nutrition Assistance Policy, Domestic Social Policy Division.

[24] For more information on Section 32, see CRS Report RL34081, *Farm and Food Support Under USDA's Section 32 Program*, by Jim Monke.

[25] Authorized by §107(a) of the Department of Defense, Emergency Supplemental Appropriations to Address Hurricanes in the Gulf of Mexico, and Pandemic Influenza Act, 2006 (P.L. 109-148), as amended and codified under 16 U.S.C. §3831a.

[26] Authorized in Section 401 of the Agricultural Credit Act of 1978 (P.L. 95-334), as amended and codified under 16 U.S.C. §§2201-2205. For more information on ECP and other land rehabilitation programs, see CRS Report R42854, *Emergency Assistance for Agricultural Land Rehabilitation*.

[27] For further information on Food and Nutrition Service's disaster relief authorities and actions, see USDA-FNS website, available at http://www.fns.usda.gov/disasters/disaster.htm.

[28] Authorized in §216 of P.L. 81-516 and §403 of the Agricultural Credit Act of 1978 (P.L. 95-334), as amended. Codified under 16 U.S.C. §2203 and 33 U.S.C. §701b-1. For more information on EWP and other land rehabilitation programs, see CRS Report R42854, *Emergency Assistance for Agricultural Land Rehabilitation*, by Megan Stubbs.

[29] This section was authored by Kelsi Bracmort, Specialist in Agricultural Conservation and Natural Resources Policy, Resources, Science, and Industry Division.

[30] The FS reports that no funds were provided to Tennessee for any of the hurricanes.

[31] Additional information provided in the FSA section of this chapter.

[32] Additional information provided in the FEMA section of this report. Region 8 encompasses 13 states including the six states identified for this request.

[33] E-mail from the Forest Service, December 10, 2012.

[34] E-mail from the Forest Service, December 11, 2012.

[35] Outlays as of December 31, 2009 were $63 million.

[36] This section was authored by Harold F. Upton, Analyst in Natural Resources Policy, Resources, Science, and Industry Division.

[37] This section was authored by Eugene Boyd, Analyst in Federalism and Economic Development Policy, Government and Finance Division.

[38] Also cited as §209(c)(2) of P.L. 89-136.

[39] Other qualifying events eligible for EAA assistance, as outlined in 42 U.S.C. §3149, include communities whose economies have been injured by military-related reductions including base closures or realignments, defense contractor reductions in force, or Department of Energy defense related funding reduction; international trade; fishery failures; or the loss of manufacturing jobs.

[40] United States Department of Commerce, Economic Development Administration, U.S. Economic Development Administration Fiscal year 2010 Annual Report, Section Four Appendix, Washington, DC, 2010, pp. 74-77, http://www.eda.gov/pdf/EDA_FY_2010_Annual_Report-APPENDIX.pdf.

[41] This section was authored by Charles Stern, Specialist in Natural Resources Policy, Resources, Science and Industry Division.

[42] This section was authored by Lawrence Kapp, Specialist in Military Manpower Policy, Foreign Affairs, Defense and Trade Division; Amy Belasco, Specialist in U.S. Defense Policy and Budgets, Foreign Affairs, Defense and Trade Division; and Dan Else, Specialist in National Defense, Foreign Affairs, Defense and Trade Division. Program summary information was taken from Department of Defense budget documents and H.Rept. 109-359, Conference Report to Accompany H.R. 2863, Department of Defense, Emergency Supplemental Appropriations to Address Hurricanes in the Gulf of Mexico, and Pandemic Influenza Act, 2006.

[43] H.Rept. 109-359, Conference Report to Accompany H.R. 2863, *Department of Defense, Emergency Supplemental Appropriations to Address Hurricanes in the Gulf of Mexico, and Pandemic Influenza Act, 2006*, p. 496.

[44] §2203, P.L. 109-234, provided that $140 million was available for infrastructure improvements to Gulf Coast shipyards damaged in 2005.

[45] This section was authored by Rebecca Skinner, Specialist in Education Policy, Domestic Social Policy Division.

[46] While not provided through education-related disaster relief legislation, Louisiana also received $20.9 million through the Charter School Program authorized under Title V-B-1 of

the Elementary and Secondary Education Act specifically to help reopen charter schools damaged by Hurricanes Katrina and Rita, help create 10 new charter schools and expand existing charter schools to accommodate displaced students. (For more information, see U.S. Department of Education, "Louisiana Awarded $20.9 Million No Child Left Behind Grant to Assist Damaged Charter Schools, Create New Charter Schools," press release, September 30, 2005, available at http://www2.ed.gov/news/pressreleases/2005/09/09302005.html).

[47] In addition to funding, P.L. 109-148 provided general waiver authority for the Secretary of Education related to maintenance of effort (MOE) requirements; the use of federal funds to supplement, not supplant non-federal funds; and matching contributions for programs administered by the Secretary. It also modified hold harmless provisions for the Elementary and Secondary Education Act (ESEA) Title I-A Grants to Local Educational Agencies program and modified highly qualified teacher provisions contained in ESEA Title I-A.

[48] Of the total appropriation for Temporary Emergency Impact Aid for Displaced Students, only $878 million was distributed, as the remaining funds were not needed by states under this program.

[49] Hawaii did not receive any funds through this program.

[50] 42 U.S.C. §11433.

[51] The eight states that received funds included Alabama, Arkansas, Florida, Georgia, Louisiana, Mississippi, Tennessee, and Texas.

[52] None of these funds were provided in response to the Gulf Coast hurricanes of 2005.

[53] While data were not available on the specific disasters experienced by the LEAs that received funding, data were available on the specific types of disasters for which institutions of higher education (IHEs) received funds under the Higher Education Disaster Relief program (P.L. 110-329), which also provided aid in response to natural disasters that occurred in 2008. According to these data, all IHEs in Louisiana that received funds were affected by Hurricane Gustav or Ike. Most IHEs in Texas that received funds were affected by Hurricane Ike. A few IHEs in Texas were affected by Hurricane Dolly, accounting for a relatively small portion of the funds allocated to IHEs in Texas. IHEs in Florida that received funding were affected by Tropical Storm Fay. LEAs in Iowa and Illinois received the remaining funds available to LEAs.

[54] The 24 states in which IHEs received funds included Alabama, Arizona, Arkansas, California, Colorado, Florida, Georgia, Illinois, Iowa, Kentucky, Louisiana, Maryland, Massachusetts, Michigan, Minnesota, Mississippi, Missouri, New Jersey, New York, Ohio, Tennessee, Texas, Utah, and Virginia.

[55] None of these funds were provided in response to the Gulf Coast hurricanes of 2005.

[56] Total obligations under this program were $15,028,360.

[57] IHEs in Arkansas, Colorado, Iowa, Illinois, Indiana, and Kentucky also received funds under this program.

[58] For a more detailed discussion of federal education-related hurricane relief, see CRS Report R42881, *Education-Related Regulatory Flexibilities, Waivers, and Federal Assistance in Response to Disasters and National Emergencies*, coordinated by Cassandria Dortch.

[59] This section was authored by Karen Lynch, Specialist in Social Policy, Domestic Social Policy Division.

[60] For additional information, see CRS Report RL30952, *Head Start: Background and Issues*, by Karen E. Lynch.

[61] See Division B of P.L. 109-148. The appropriations language specified that costs of renovations may be covered "to the extent reimbursements from FEMA and insurance companies do not fully cover such costs."

[62] U.S. Department of Health and Human Services, Administration for Children and Families, "FY2008 Justification of Estimates for Appropriations Committees," February 2007, p. 91.

[63] For additional information, see CRS Report 94-953, *Social Services Block Grant: Background and Funding*, by Karen E. Lynch.

[64] See Division B of P.L. 109-148.

[65] See Division B of P.L. 110-329.

[66] For the purpose of allocating these funds, ACF counted major disasters occurring between January and September of 2008 for which FEMA Individual Assistance was authorized, plus Hurricanes Katrina and Rita.

[67] Of the $944 million, $519 million came from funds appropriated in P.L. 109-148, while $425 million came from funds appropriated in P.L. 110-329. Notably, allocations from the latter appropriation were developed based on needs resulting from a broader array of storms. In addition to accounting for Hurricanes Katrina, Rita, Gustav, and Ike, the formula for allocating these funds also took into account other major disasters of CY2008 that qualified for the FEMA Individual Assistance program, such as Tropical Storm Fay in Florida, Hurricane Dolly in Texas, and various other severe storms, tornados, and floods. For state-by-state allocation and expenditure data for these supplemental appropriations, see CRS Report 94-953, *Social Services Block Grant: Background and Funding*, by Karen E. Lynch.

[68] See §2002(c) of Title XX-A of the Social Security Act.

[69] The expenditure deadline for the $550 million in supplemental SSBG funds appropriated in P.L. 109-148 was initially September 30, 2007. This deadline was extended, by P.L. 110-28, through September 30, 2009. The expenditure deadline for the $600 million in supplemental SSBG funds appropriated in P.L. 110-329 was initially September 30, 2010. This deadline was extended, by P.L. 111-285, through September 30, 2011.

[70] These data were current as of December 15, 2011, and should not be considered final. Terms and conditions of SSBG grant awards give states an additional 90 days (in this case, until December 30, 2011) to liquidate funds that had already been obligated at the end of the fiscal year. Final expenditure data have not yet been made available.

[71] 45 C.F.R. §96.74(b).

[72] U.S. Department of Health and Human Services, Administration for Children and Families, "Social Services Block Grant Program Annual Report 2009, Chapter 5," available at http://archive.acf.hhs.gov/programs/ocs/ssbg/reports/ 2009/index.html.

[73] This section was authored by Sarah A. Lister, Specialist in Public Health and Epidemiology, Domestic Social Policy Division.

[74] CRS Report RL34758, *The National Response Framework: Overview and Possible Issues for Congress*, by Bruce R. Lindsay.

[75] Ibid, and CRS Report R40708, *Disaster Relief Funding and Emergency Supplemental Appropriations*, by Bruce R. Lindsay and Justin Murray.

[76] P.L. 109-171, Deficit Reduction Act (DRA), §6201, 120 Stat. 132-134, February 8, 2006; and P.L. 109-62, Second Emergency Supplemental Appropriations Act to Meet Immediate Needs Arising From the Consequences of Hurricane Katrina, 2005, 119 Stat. 1991, September 8, 2005. The $2 billion appropriated in the DRA was in addition to $100 million appropriated earlier to the National Disaster Medical System, some of which was also transferred for this purpose. See GAO, *Hurricane Katrina: Allocation and Use of $2 Billion for Medicaid and Other Health Care Needs*, GAO-07-67, February 28, 2007, and GAO,

*Hurricane Katrina: CMS and HRSA Assistance to Sustain Primary Care Gains in the Greater New Orleans Area*, GAO-10-773R, June 30, 2010.

[77] Congress also provided $90 million in grants to states for high-risk pools that provide health insurance to individuals who are otherwise uninsurable. P.L. 109-171 (DRA), §6202, 120 Stat. 134. Almost all states were eligible and received awards under this program. Although it was not the primary focus, some states may have used the funds to provide insurance coverage to hurricane evacuees.

[78] P.L. 109-234, Emergency Supplemental Appropriations Act for Defense, the Global War on Terror, and Hurricane Recovery, 2006, 120 Stat. 463, June 15, 2006. A portion of funds for communications equipment was provided to North Carolina, which deployed a field hospital to the Gulf Coast; funding was used to facilitate that aid.

[79] HHS, "HHS Provides Prescription Drug and Durable Medical Equipment Assistance for Uninsured Texas Victims of Hurricane Ike," press release, September 12, 2008; HHS, "HHS Awards Grants to Support Minority Health," press release, September 30, 2005; and HHS, "HHS Awards $600,000 in Emergency Mental Health Grants to Four States Devastated by Hurricane Katrina," press release, September 13, 2005, available at http://www.hhs.gov/news.

[80] These waiver authorities are described in CRS Report R40560, *The 2009 Influenza Pandemic: Selected Legal Issues*, coordinated by Kathleen S. Swendiman and Nancy Lee Jones.

[81] No-year funds are available until they are expended.

[82] See "Federal Funding to Support an ESF-8 Response," in CRS Report RL33579, *The Public Health and Medical Response to Disasters: Federal Authority and Funding*, by Sarah A. Lister.

[83] This section was authored by Bruce R. Lindsay, Analyst in American National Government, Government and Finance Division.

[84] 42 U.S.C. §5121 et seq. For further analysis on the Stafford Act, see CRS Report RL33053, *Federal Stafford Act Disaster Assistance: Presidential Declarations, Eligible Activities, and Funding*, by Francis X. McCarthy.

[85] 42 U.S.C. §5170b(a)(1).

[86] 42 U.S.C. §5147.

[87] Federal Emergency Management Agency, "About FEMA," October 31, 2012, available at http://www.fema.gov/about/index.shtm#0.

[88] The DRF is the main account used to fund a wide variety of programs, grants, and other forms of emergency and disaster assistance to states, local governments, certain nonprofit entities, and families and individuals affected by disasters. In most cases, funding from the DRF is released after the President has issued a declaration pursuant to the Stafford Act. For further analysis on the DRF, see CRS Report R40708, *Disaster Relief Funding and Emergency Supplemental Appropriations*, by Bruce R. Lindsay and Justin Murray. For further analysis on declaration process, see CRS Report RL34146, *FEMA's Disaster Declaration Process: A Primer*, by Francis X. McCarthy.

[89] §406, 42 U.S.C. §5172.

[90] §408, 42 U.S.C. §5174.

[91] §407, 42 U.S.C. §5173.

[92] §404, 42 U.S.C. §5170c.

[93] §403, 42 U.S.C. §5170b.

[94] Making continuing appropriations for fiscal year 2013, and for other purposes.

[95] Department of Homeland Security, "Agency Information Collection Activities: Proposed Collection; Comment Request," 69 *Federal Register* 9350, February 27, 2004.

[96] This section was authored by Jared T. Brown, Analyst in Emergency Management and Homeland Security Policy, Government and Finance Division. For more on the CDL program, see CRS Report R42527, *FEMA's Community Disaster Loan Program: History, Analysis, and Issues for Congress*, by Jared T. Brown.

[97] P.L. 93-288, as amended; 42 U.S.C. §5121 et seq. The program was first authorized by the Disaster Relief Act of 1974. The 1974 Act was renamed the Stafford Act by the Disaster Relief and Emergency Assistance Amendments of 1988, P.L. 100-707, §102. Prior to the Act in 1988, the CDL Program was codified as §414, not §417.

[98] 42 U.S.C. §5184(a).

[99] 6 U.S.C. §314(a)(8).

[100] U.S. Congress, Senate Committee on Public Works, *Disaster Relief Act Amendments of 1974*, Report to Accompany S. 3062, 93rd Cong., 2nd sess., April 9, 1974, S.Rept. 93-778, p. 9.

[101] As defined in 42 U.S.C. §5122 of the Stafford Act, the definition of "local government" is:
(A) a county, municipality, city, town, township, local public authority, school district, special district, intrastate district, council of governments (regardless of whether the council of governments is incorporated as a nonprofit corporation under state law), regional or interstate government entity, or agency or instrumentality of a local government;
(B) an Indian tribe or authorized tribal organization, or Alaska Native village or organization; and
(C) a rural community, unincorporated town or village, or other public entity, for which an application for assistance is made by a state or political subdivision of a state.

[102] 42 U.S.C. §5184.

[103] 44 C.F.R. §206.366(d)(1).

[104] The regulations for traditional CDLs are found in 44 C.F.R. §§206.360-367. The regulations for the special CDLs are found in 44 C.F.R. §§206.370-377.

[105] For a full list of major disaster declarations during 2008, see http://www.fema.gov/news/disasters.fema?year=2008.

[106] In most appropriation bills, the program will be identified either by this account or as §417 of the Stafford Act.

[107] 42 U.S.C. §5162.

[108] The CDL program is subject to the Federal Credit Reform Act of 1990, as amended (P.L. 101-508, FCRA). The FCRA changed the accounting method for measuring the cost of federal direct loans and loan guarantees, starting in FY1992. Under the FCRA, discretionary programs providing new direct loan obligations or new loan guarantee commitments require appropriations of budget authority equal to the loans' estimated subsidy costs. Furthermore, the appropriations bill must include an estimate for the dollar amount of the new direct loan obligations that are supported by the subsidy budget authority appropriated to the agency for its credit program. The subsidy rate for any loan program is calculated by CBO in each appropriation. Therefore, appropriations to the DADLP account for the purposes of the CDL program have varied in the total dollar amount of loans that can be issued per appropriated dollar.

[109] In addition, one could consider the interest rate as being "subsidized" because the market may charge a higher rate of interest on the loan than the federal government. If it is the case that the interest rate is subsidized, one should also account for the subsidy in any hypothetical calculation of the "cost" of the program to the federal taxpayer. The CDL statute is silent on what interest rate should be charged for loans issued under the program. See 44 C.F.R. §206.361(c).

[110] Current information on the loan cancellation process for each of these disasters can be obtained by contacting CRS analyst Jared T. Brown, jbrown@crs.loc.gov, 7-4918.

[111] This section was authored by Maggie McCarty, Specialist in Housing Policy and Eugene Boyd, Analyst in Federalism and Economic Development Policy.

[112] 42 U.S.C. §5321. For funds designated under this chapter by a recipient to address the damage in an area for which the President has declared a disaster under title IV of the Robert T. Stafford Disaster Relief and Emergency Assistance Act [42 U.S.C. §5170 et seq.], the Secretary may suspend all requirements for purposes of assistance under §5306 of this title for that area, except for those related to public notice of funding availability, nondiscrimination, fair housing, labor standards, environmental standards, and requirements that activities benefit persons of low- and moderate-income.

[113] The Housing Authority of New Orleans sustained so much damage as a result of the storm that they contracted with a PHA in Harris County, Texas to administer their voucher program for them.

[114] For more information about the Shelter Plus Care program, see CRS Report RL33764, *The HUD Homeless Assistance Grants: Programs Authorized by the HEARTH Act*, by Libby Perl; for more information about Section 8 vouchers, see CRS Report RL32284, *An Overview of the Section 8 Housing Programs: Housing Choice Vouchers and Project-Based Rental Assistance*, by Maggie McCarty.

[115] This section was authored by Nathan James, Analyst in Crime Policy, Domestic Social Policy Division.

[116] 28 U.S.C. §501.

[117] U.S. Department of Justice, Offices of the United States Attorneys, "United States Attorneys' Mission Statement," available at http://www.justice.gov/usao/about/mission.html.

[118] U.S. Congress, House, *Making Emergency Supplemental Appropriations for the Fiscal Year Ending September 30, 2006 and Other Purposes*, Conference Report, 109th Cong., 2nd sess., June 8, 2006, H.Rept. 109-494 (Washington: GPO, 2006), p. 128.

[119] U.S. Congress, House, *Making Appropriations for the Department of Defense for the Fiscal Year Ending September 30, 2006, and Other Purposes*, Conference Report, 109th Cong., 1st sess., December 18, 2005, H.Rept. 109-359 (Washington: GPO, 2005), p. 514.

[120] U.S. Congress, House, *Making Emergency Supplemental Appropriations for the Fiscal Year Ending September 30, 2006 and Other Purposes*, Conference Report, 109th Cong., 2nd sess., June 8, 2006, H.Rept. 109-494 (Washington: GPO, 2006), p. 128.

[121] U.S. Department of Justice, U.S. Marshals Service, "Overview of the U.S. Marshals Service," available at http://www.usmarshals.gov/duties/factsheets/general-2011.html.

[122] Ibid.
[123] Ibid.
[124] Ibid.
[125] Ibid.
[126] Ibid.
[127] Ibid.
[128] Ibid.

[129] U.S. Department of Justice, "Assets Forfeiture Program," available at http://www.justice.gov/jmd/afp/.

[130] U.S. Department of Justice, "U.S. Marshals Service, Overview of the U.S. Marshals Service," available at http://www.usmarshals.gov/duties/factsheets/general-2011.html.

[131] Ibid.
[132] Ibid.

[133] U.S. Congress, House, *Making Appropriations for the Department of Defense for the Fiscal Year Ending September 30, 2006, and Other Purposes*, Conference Report, 109th Cong., 1st sess., December 18, 2005, H.Rept. 109-359 (Washington: GPO, 2005), p. 514.
[134] U.S. Department of Justice, Federal Bureau of Investigation, "Quick Facts," available at http://www.fbi.gov/aboutus/quick-facts.
[135] Ibid.
[136] Ibid.
[137] Ibid.
[138] Ibid.
[139] Ibid.
[140] U.S. Department of Justice, Federal Bureau of Investigation, "Uniform Crime Reports," available at http://www.fbi.gov/about-us/cjis/ucr.
[141] U.S. Department of Justice, Federal Bureau of Investigation, "Combined DNA Index System (CODIS)," available at http://www.fbi.gov/about-us/lab/biometric-analysis/codis.
[142] U.S. Department of Justice, Federal Bureau of Investigation, "N-Dex: National Law Enforcement Data Exchange," available at http://www.fbi.gov/about-us/cjis/n-dex.
[143] U.S. Department of Justice, Federal Bureau of Investigation, "Integrated Automated Fingerprint Identification System," available at http://www.fbi.gov/about-us/cjis/fingerprints_biometrics/iafis.
[144] U.S. Department of Justice, Federal Bureau of Investigation, "National Instant Criminal Background Check System," available at http://www.fbi.gov/about-us/cjis/nics.
[145] U.S. Department of Justice, Federal Bureau of Investigation, "National Crime Information Center," available at http://www.fbi.gov/about-us/cjis/ncic.
[146] U.S. Congress, House, *Making Appropriations for the Department of Defense for the Fiscal Year Ending September 30, 2006, and Other Purposes*, Conference Report, 109th Cong., 1st sess., December 18, 2005, H.Rept. 109-359 (Washington: GPO, 2005), p. 514.
[147] U.S. Department of Justice, Drug Enforcement Administration, "DEA History," available at http://www.justice.gov/dea/about/history.shtml.
[148] U.S. Department of Justice, Drug Enforcement Administration, "Domestic Office Locations," available at http://www.justice.gov/dea/about/Domesticoffices.shtml. U.S. Department of Justice, Drug Enforcement Administration, Foreign Office Locations, available at http://www.justice.gov/dea/about/foreignoffices.shtml.
[149] U.S. Department of Justice, Drug Enforcement Administration, "DEA Mission Statement," available at http://www.justice.gov/dea/about/mission.shtml.
[150] Ibid.
[151] U.S. Congress, House, *Making Appropriations for the Department of Defense for the Fiscal Year Ending September 30, 2006, and Other Purposes*, Conference Report, 109th Cong., 1st sess., December 18, 2005, H.Rept. 109-359 (Washington: GPO, 2005), p. 514.
[152] CRS Report R41206, *The Bureau of Alcohol, Tobacco, Firearms and Explosives (ATF): Budget and Operations for FY2011*, by William J. Krouse.
[153] U.S. Department of Justice, Bureau of Alcohol, Tobacco, Firearms and Explosives, "ATF's History," available at http://www.atf.gov/about/history/.
[154] U.S. Department of Justice, Bureau of Alcohol, Tobacco, Firearms and Explosives, Bureau of Alcohol, Tobacco, Firearms and Explosives, "Congressional Budget Submission, Fiscal Year 2013," p. 1, available at http://www.justice.gov/jmd/2013justification/pdf/fy13-atf-justification.pdf.

[155] U.S. Congress, House, *Making Appropriations for the Department of Defense for the Fiscal Year Ending September 30, 2006, and Other Purposes*, Conference Report, 109th Cong., 1st sess., December 18, 2005, H.Rept. 109-359 (Washington: GPO, 2005), p. 515.
[156] U.S. Department of Justice, Bureau of Prisons, "About the Bureau of Prisons," available at http://www.bop.gov/about/index.jsp.
[157] U.S. Department of Justice, Bureau of Prisons, "Mission and Vision of the Bureau of Prisons," available at http://www.bop.gov/about/mission.jsp.
[158] U.S. Department of Justice, Bureau of Prisons, "About the Bureau of Prisons," available at http://www.bop.gov/about/index.jsp.
[159] U.S. Congress, House, *Making Appropriations for the Department of Defense for the Fiscal Year Ending September 30, 2006, and Other Purposes*, Conference Report, 109th Cong., 1st sess., December 18, 2005, H.Rept. 109-359 (Washington: GPO, 2005), p. 515.
[160] U.S. Department of Justice, Office of Justice Programs, "About Us," available at http://www.ojp.gov/about/about.htm.
[161] U.S. Department of Justice, Office of Justice Programs, Office of Justice Programs, "Congressional Budget Submission, Fiscal Year 2013," pp. 19-20, available at http://www.justice.gov/jmd/2013justification/pdf/fy13-ojpjustification.pdf.
[162] Ibid.
[163] Ibid., p. 19.
[164] Ibid.
[165] Ibid., p. 20.
[166] Ibid.
[167] Ibid.
[168] Ibid.
[169] Ibid. p. 19.
[170] Ibid.
[171] This section was authored by David Bradley, Specialist in Labor Economics, Domestic Social Policy Division.
[172] U.S. Department of Labor, Employment and Training Administration, "About ETA", available at http://www.doleta.gov/etainfo/.
[173] As ETA notes, "The primary purpose of a disaster project is to create temporary employment to assist with clean-up activities. The initial award will restrict the clean-up period to 6 months from the date of grant award, until there is a subsequent modification (e.g., fully documented plan or other request) that justifies a longer clean-up period." ETA, "Eligible Events for National Emergency Grant Funding," available at http://www.doleta.gov/neg/dislocation.cfm#3.
[174] Specifically, WIA §132(a)(2)(A) and (a)(2)(B) require that 20% of the amount appropriated for Dislocated Worker Employment and Training Activities be reserved for national emergency grants, projects, and technical assistance. The remaining 80% is to be used for state formula grants.
[175] NEG award amounts were obtained from the ETA Office of National Response (ONR). ONR reports grants awarded by state and type of project in each calendar year. Because the supplemental appropriations became law December 30, 2005 (P.L. 109-148), the amounts reported in Table 26 are for calendar year 2006 only. It should be noted that additional NEG funding was provided to these five states in other calendar years. Florida, for example, received $8.5 million in NEG funding in 2005 for hurricane-related emergencies; however, given the timing of P.L. 109-148, it does not appear that Florida's funding came from the supplemental appropriations identified in Table 26.

[176] The Job Corps program is scheduled to open its 126th center in Manchester, NH, in 2013. In addition, there is a new center under construction in Wyoming. See U.S. Department of Labor, "FY2013 Budget Justification of Appropriation Estimates for Committee on Appropriations, Vol. I," p. OJC-28.

[177] The PY (program year) 2005 Job Corps Annual Report indicated that, "Following a recent appropriation from Congress, Job Corps is on the fast track to restoring the Gulfport and New Orleans Job Corps centers, which sustained damage during Hurricane Katrina." See U.S. Department of Labor, *Job Corps Annual Report: Program Year July 1, 2005 - June 30, 2006*, Washington, DC, 2006, p. 27, available at http://www.jobcorps.gov/Libraries/pdf/py05report.sflb.

[178] This section was authored by Robert S. Kirk, Specialist in Transportation Policy, Resources, Science, and Industry Division.

[179] Regulatory Reference: 23 C.F.R. Part 668.

[180] CRS Report R42804, *Emergency Relief Program: Federal-Aid Highway Assistance for Disaster-Damaged Roads and Bridges*, by Robert S. Kirk.

[181] This total includes $1 million in Airport Improvement Program funding provided on September 19, 2008.

[182] The FAA was the lead Operational Administration for the Katrina disaster mission assignment responses. Most of the mission assignment costs overseen by FAA following Katrina were for services provided by Landstar Express America, Inc. Landstar provided transport services for the air, sea, and land transportation of supplies and resources.

[183] This section was authored by Sidath Viranga Panangala, Specialist in Veterans Policy, Domestic Social Policy Division.

[184] For more information see CRS Report RL33409, *Veterans' Medical Care: FY2007 Appropriations*, by Sidath Viranga Panangala.

[185] U.S. Congress, House Committee on Veterans' Affairs, *Deconstructing the U.S. Department of Veterans Affairs Construction Planning*, 112th Cong., 1st sess., April 5, 2011 (Washington: GPO, 2011), p.73 and p. 69.

[186] Department of Veteran Affairs, Southeast Louisiana Veterans Health Care System, "Project Legacy - Frequently Asked Questions," February 21, available at http://www.neworleans.va.gov/ Project_Legacy_Frequently_Asked_Questions.asp.

[187] This section was authored by Ann Lordeman, Specialist in Social Policy, Domestic Social Policy Division.

[188] Information on the use of the $10 million appropriated under P.L. 109-234 was provided by the Corporation for National and Community Service in correspondence with CRS on July 15, 2009.

[189] This section was authored by Robert Esworthy, Environmental Specialist, Resources, Science and Industry Division.

[190] See CRS Report RL30798, *Environmental Laws: Summaries of Major Statutes Administered by the Environmental Protection Agency*, coordinated by David M. Bearden. See also U.S. EPA, "Emergency Management: Laws Defining EPA's Emergency Management Program," available at http://www.epa.gov/oem/lawsregs.htm.

[191] See U.S. EPA, "Response to 2005 Hurricanes," available at http://www.epa.gov/katrina/. See also CRS Report RL33115, *Cleanup After Hurricane Katrina: Environmental Considerations*, by Robert Esworthy et al. For information regarding EPA's actions supporting FEMA and working closely with federal agencies and the states in response to Hurricane Sandy of 2012, see U.S. EPA, "Hurricane Sandy Response and Recovery," available at http://www.epa.gov/ sandy/.

[192] 33 U.S.C. §2701 et. seq., and 33 U.S.C. §1321, respectively. For further discussion of the authorities of OPA and Section 311 of the Clean Water Act, see CRS Report RL33705, *Oil Spills in U.S. Coastal Waters: Background and Governance*, by Jonathan L. Ramseur.

[193] The term "environment" includes surface and subsurface lands, surface waters, groundwater, and ambient air, making the response authorities for hazardous substances broader in terms of their physical reach than that for oil spills. 42 U.S.C. §9601 et. seq. For further discussion of the authorities of CERCLA, see CRS Report R41039, *Comprehensive Environmental Response, Compensation, and Liability Act: A Summary of Superfund Cleanup Authorities and Related Provisions of the Act*, by David M. Bearden.

[194] Executive Order 12580 delegated the President's authorities under CERCLA, and Executive Order 12777 delegated the President's authorities under OPA and Section 311 of the Clean Water Act. Executive Order 13286 amended these executive orders to reflect the transfer of the U.S. Coast Guard from the Department of Transportation to the Department of Homeland Security in 2003.

[195] Subtitle I of the Solid Waste Disposal Act addresses petroleum leaked from underground storage tanks. This role is performed mainly by the states under cooperative agreements with EPA.

[196] For information on the National Infrastructure Protection Plan and sector-specific agency roles, see the Department of Homeland Security's website, available at http://www.dhs.gov/national-infrastructure-protection-plan.

[197] 42 U.S.C. §300i.

[198] 40 CFR Part 300.

[199] EPA is the primary agency for the inland zone and incidents affecting both inland and coastal zones; the U.S. Coast Guard has primary responsibility for coastal incidents and often acts as co-lead.

[200] For more information about EPA responsibilities under the National Response Framework, including those under individual ESFs, see EPA's "Federal Response Plans" website, available at http://www.epa.gov/ homelandsecurityportal/laws-fedresponse.htm#nrf.

[201] For more information, see CRS Report RL34576, *Managing Disaster Debris: Overview of Regulatory Requirements, Agency Roles, and Selected Challenges*, by Linda Luther.

[202] See EPA's website: "Fuels Waivers Response to 2005 Hurricanes" available at http://www.epa.gov/compliance/ katrina/waiver/index.html.

[203] EPA activities included assessment and containment of existing Superfund sites and releases from underground storage tanks. EPA uses funds from the Superfund appropriations account to pay for emergency response activities for all pre-existing Superfund sites; see "Policy Guidance on ESF #10 Mission Assignments," available at http://coop.fema.gov/government/grant/pa/9523_8b.

[204] The funding received includes $800,000 received through a U.S. Army Corp of Engineers interagency agreement. The total received amounts as reported by EPA reflect adjustments resulting from quarterly reviews on all Mission Assignments and Interagency Agreements performed jointly by FEMA/DHS, EPA Cincinnati finance office, EPA Regional Program Office, and Federal Coordinating Officers, and funding EPA provided back to FEMA.

[205] See U.S. EPA publication "Public Assistance for Water and Wastewater Utilities in Emergencies and Disasters," EPA 817-F-10-009, Office of Water, Aug, 2010, available at http://water.epa.gov/infrastructure/watersecurity/ emerplan/upload/Public-Assistance-for-Water-and-Wastewater-Utilities-in-Emergencies-and-Disasters.pdf. See also FEMA 322 Public Assistance Guide, under Category F, and "Federal Funding for Utilities -

Water/Wastewater - in National Disasters (Fed FUNDS)" available at http://water.epa.gov/infrastructure/watersecurity/funding/fedfunds/ index.cfm.

[206] EPA must allocate the Clean Water SRF grants among the states according to a formula specified in the Clean Water Act itself, whereas the Safe Drinking Water Act authorizes EPA to develop a formula for allocating the Drinking Water SRF grants among the states that is to reflect the proportional share of each state's funding needs.

[207] Includes $438.4 million allotted to these states in FY2009 from the total $ 4.0 billion in CWSRF supplemental appropriations included in the American Recovery and Reinvestment Act of 2009 (ARRA, P.L. 111-5)

[208] Includes $315.4 million allotted to these states in FY2009 from the total $2.0 billion DWSRF supplemental appropriations included in the ARRA (P.L. 111-5).

[209] This section was authored by Matthew Eric Glassman, Analyst in American National Government, Government and Finance Division.

[210] The 94 U.S. judicial districts are organized into 12 regional circuits, each of which has a United States court of appeals. A court of appeals hears appeals from the district courts located within its circuit, as well as appeals from decisions of federal administrative agencies.

[211] P.L. 109-148, Department of Defense, Emergency Supplemental Appropriations to Address Hurricanes in the Gulf of Mexico, and Pandemic Influenza Act, 2006.

[212] This section was authored by Bruce R. Lindsay, Analyst in American National Government, Government and Finance Division.

[213] U.S. Small Business Administration, "Disaster Assistance," available at http://www.sba.gov/services/ disasterassistance/.

[214] This section was authored by Francis X. McCarthy, Analyst in Emergency Management Policy, Government and Finance Division.

[215] For more information on cost-shares see CRS Report R41101, *FEMA Disaster Cost-Shares: Evolution and Analysis*, by Francis X. McCarthy.

[216] P.L. 109-234, 115 Stat. 671.

[217] This section was authored by Bruce R. Lindsay, Analyst in American National Government, Government and Finance Division.

[218] U.S. Department of Homeland Security/Federal Emergency Management Agency, "Declared Disasters by Year of State," available at http://www.fema.gov/news/disaster_totals_annual.fema.

[219] Funding information is currently provided in different formats including obligations, allocations, and expenditures.

[220] For historical information on major disaster declarations see CRS Report R42702, *Stafford Act Declarations 1953- 2011: Trends and Analyses, and Implications for Congress*, by Bruce R. Lindsay and Francis X. McCarthy.

[221] For more information on the DRF and supplemental appropriations see CRS Report R40708, *Disaster Relief Funding and Emergency Supplemental Appropriations*, by Bruce R. Lindsay and Justin Murray.

[222] The BCA uses a 10-year average of major disaster costs under the Stafford Act to limit spending on disasters. For more information on the BCA see CRS Report R41965, *The Budget Control Act of 2011*, by Bill Heniff Jr., Elizabeth Rybicki, and Shannon M. Mahan, and CRS Report R42352, *An Examination of Federal Disaster Relief Under the Budget Control Act*, by Bruce R. Lindsay, William L. Painter, and Francis X. McCarthy.

[223] For more information on offsets and supplemental appropriations see CRS Report R42458, *Offsets, Supplemental Appropriations, and the Disaster Relief Fund: FY1990-FY2013*, by William L. Painter.
[224] This section was added in full committee markup of the legislation. For a more in-depth discussion of procedural considerations for offsetting amendments, see CRS Report RL31055, *House Offset Amendments to Appropriations Bills: Procedural Considerations*, by Jessica Tollestrup.
[225] P.L. 93-288, 42 U.S.C. §5122.
[226] Richard T. Sylves, Disaster Policy and Politics: Emergency Management and Homeland Security (Washington, DC: CQ Press, 2008), p. 79.
[227] Department of Homeland Security, Office of Inspector General, Opportunities to Improve FEMA's Public Assistance Preliminary Damage Assessment Process, pp. 5-7.
[228] U.S. Government Accountability Office, DISASTER ASSISTANCE; Improvement Needed in Disaster Declaration Criteria and Eligibility Assurance Procedures, GAO-01-837, August, 2001, p.11
[229] For example, S. 1630, the Disaster Recovery Act of 2011, which was introduced on September 23, 2011, and referred to the Committee on Homeland Security and Governmental Affairs, would have amended the Stafford Act to authorize the President to declare a catastrophic incident if a recommendation was issued by an independent panel of experts.
[230] Assistance for emergency declarations is capped at $5 billion per incident.
[231] 42 U.S.C. §5170c.
[232] For additional discussion on this topic see CRS Report R41101, *FEMA Disaster Cost-Shares: Evolution and Analysis*, by Francis X. McCarthy.
[233] Ibid.
[234] There is however, a limit on how much any one household can receive ($31,400 at the time of this report).

In: Federal Disaster Assistance ...
Editor: Madeline Payton
ISBN: 978-1-63117-886-3
© 2014 Nova Science Publishers, Inc.

*Chapter 2*

# DISASTER RELIEF FUNDING AND SUPPLEMENTAL APPROPRIATIONS FOR DISASTER RELIEF[*]

*Bruce R. Lindsay and Justin Murray*

## SUMMARY

When a state is overwhelmed by an emergency or disaster, the governor may request assistance from the federal government. Federal assistance is contingent on whether the President issues an emergency or major disaster declaration.

Once the declaration has been issued the Federal Emergency Management Agency (FEMA) provides disaster relief through the use of the Disaster Relief Fund (DRF), which is the source of funding for the Robert T. Stafford Emergency Relief and Disaster Assistance Act response and recovery programs. Congress appropriates money to the DRF to ensure that funding for disaster relief is available to help individuals and communities stricken by emergencies and major disasters (in addition, Congress appropriates disaster funds to other accounts administered by other federal agencies pursuant to federal statutes that authorize specific types of disaster relief).

---

[*] This is an edited, reformatted and augmented version of a Congressional Research Service publication, CRS Report for Congress R40708, dated August 5, 2013.

Historically, the DRF is generally funded at a level that is sufficient for what are known as "normal" disasters. These are incidents for which DRF outlays are less than $500 million. When a large disaster occurs, funding for the DRF may be augmented through emergency supplemental appropriations. A supplemental appropriation generally provides additional budget authority during the current fiscal year to (1) finance activities not provided for in the regular appropriation; or (2) provide funds when the regular appropriation is deemed insufficient.

This methodology used to budget the DRF appears to have been altered since the passage of the Budget Control Act (P.L. 112-25). The Budget Control Act includes a series of provisions that directed the Office of Management and Budget (OMB) to annually calculate an "allowable adjustment" for disaster relief to the BCA's discretionary spending caps. That adjustment, if used, would make additional budget authority available for the federal costs incurred by major disasters declared under the Stafford Act beyond what is allowed in the regular discretionary budget allocation. The OMB calculation may have provided a mechanism that encourages a larger regular appropriation to the DRF. It is possible that larger DRF appropriations may reduce the need for supplemental appropriations.

Budgeting for disaster relief has been the subject of a great deal of debate. Some argue that more money should be appropriated in FEMA's DRF account in annual appropriations, while others maintain that augmenting the DRF through supplemental appropriations is preferable because it allows Congress to react directly to a particular situation. Others may argue that emergency supplemental appropriations are preferable for fiscal management reasons because an appropriation is not requested unless there is a real need for supplemental funding. Another argument is to revamp the budgetary process to fund disaster relief.

This report describes the various components of the DRF, including (1) what authorities have shaped it over the years; (2) how FEMA determines the amount of the appropriation requested to Congress (pertaining to the DRF); and (3) how emergency supplemental appropriations are requested. Information is also provided on funds appropriated in supplemental appropriations legislation to agencies other than the Department of Homeland Security (DHS). Aspects of debate concerning how disaster relief is budgeted are also highlighted and examined, and alternative budgetary options are summarized.

## DISASTER RELIEF FUND

The Disaster Relief Fund (DRF), sometimes referred to as the President's Disaster Relief Fund, is managed by the Federal Emergency Management Agency (FEMA). The DRF is the main account used to fund a wide variety of programs that provide grants and other support to assist state and local governments and certain nonprofit entities during disaster recovery. In most cases, funding from the DRF is released after the President has issued a declaration pursuant to the Robert T. Stafford Relief and Emergency Assistance Act (Stafford Act).[1] There are, however, some activities that may be funded by the DRF without a presidential declaration, mainly those supported by the Disaster Readiness and Support Account. The Disaster Readiness and Support Account pays for FEMA's phone centers, finance centers, and housing inspectors. Through this account certain recovery elements are already in place when the President issues a declaration.

The declaration process has the following general structure:

- the local government responds to an event;
- if the event overwhelms the local government, the state is called upon for assistance;
- a Preliminary Damage Assessment (PDA) is completed by local, state, federal, and volunteer organizations to determine loss and recovery needs;
- based on the PDA findings, a declaration request is submitted to the President byt he governor;[2]
- FEMA evaluates the governor's request and recommends action to the White House, based on the nature of the event and an assessment of the state's ability to recover; and
- the President may either approve the request or require FEMA to inform the governor the request has been denied.[3]

A President's declaration triggers the allocation of funds from the DRF, and the funding may be distributed from any one, or any combination, of three categories of disaster aid:

1. **Individual Assistance.** Individual Assistance (IA) includes disaster housing for displaced individuals, grants for needs not covered by insurance, crisis counseling, and disaster-related unemployment assistance.

2. **Public Assistance.** Public assistance (PA), which is FEMA's largest funded program, helps communities absorb the costs of emergency measures such as removing debris and repairing or replacing structures such as public buildings, roads, bridges, and utilities.
3. **Hazard Mitigation.** FEMA funds mitigation measures to prevent or lessen the effects of a future disaster through the Hazard Mitigation Grant Program.[4]

Even if the President issues an emergency or major disaster declaration, not all persons or entities affected by a disaster are eligible for disaster assistance. FEMA officials determine the need for assistance after a declaration is issued. Aid is provided only to those persons or entities determined by FEMA to need the assistance. An example of ineligibility would be a household that has access to alternative housing.[5]

## PAST AND PRESENT AUTHORITIES RELATED TO THE DISASTER RELIEF FUND

The Stafford Act authorizes appropriations to carry out disaster relief activities, but does not explicitly designate the DRF as the account for such funds. Rather, the DRF is the product of legislation and federal policies that can be traced to the post-World War II era. Prior to that time, disaster response activities were funded primarily through local efforts and voluntary groups. In cases where the federal government did offer assistance, the needs of disaster victims and affected communities were funded on an as-needed basis through appropriations that were then allocated, pursuant to the legislation, by executive branch administrators and, ultimately, the President.[6]

Arguably, the forerunners to the DRF were appropriations known as "emergency funds for the President." For example, in a 1948 appropriation, Congress allocated $500,000 to "supplement the efforts and available resources of state and local governments or other agencies, whenever ... any flood, fire, hurricane, earthquake, or other catastrophe in any part of the United States is of sufficient severity and magnitude to warrant emergency assistance by the federal government."[7] In subsequent years Congress continued to appropriate funds for disaster relief and eventually a series of authorizations were provided in Section 606 of the Disaster Relief Act of 1974 as amended.[8] These authorizations provided funding for disaster-related activities until a specified date. In 1988 Congress repealed Section 606 of the Disaster Relief Act ending extensions of authorizations.[9] Since then funds have been appropriated, though not specifically authorized.

## Public Laws Influencing the Administration of Disaster Relief

The approach to disaster relief changed from 1950 to 1979, transitioning from a largely uncoordinated and decentralized system of relief funding to one dominated by the federal government. The transition was due in part to an increased awareness of the complex responsibilities of federal and nonfederal entities before and after a disaster, and to the efforts of policymakers and administrators to address a variety of concerns—most notably civil defense.[10]

After World War II, concern over the possible use of atomic weapons and growing hostility between the United States and the Soviet Union gave rise to the Cold War. As a consequence, disaster management in the United States was organized around two tracks: (1) the threat of a nuclear war and (2) natural disasters. Several landmark federal disaster laws and policies originate from attempts by lawmakers during this era to prepare the civilian population for a potential atomic attack and provide aid after a natural disaster.[11] The most notable of these laws were the Civil Defense Act of 1950[12] and the Federal Disaster Relief Act of 1950.[13] These laws set into motion federal-to-state assistance, prompting the need for an account to fund disaster and emergency activities.

Once a framework of federal to state disaster assistance had been constructed in the 1950 statute, the process of administering disaster relief was further shaped by the Disaster Relief Acts of 1966[14] and 1974,[15] and the Robert T. Stafford Disaster Relief and Emergency Assistance Act of 1988.[16] Prompted by large disasters such as Hurricanes Betsy in 1965 and Agnes in 1972, these laws significantly increased federal involvement in emergency management, which in turn created the need for a funding mechanism to pay for emergency and disaster activities. Summaries of the statutes follow.

### *The Federal Disaster Relief Act of 1950 (P.L. 81-875)*

The Federal Disaster Relief Act established much of the framework through which disaster policy is carried out in the United States. The Federal Disaster Relief Act provided an orderly, ongoing continued means of federal assistance to states and localities to alleviate suffering and damage that resulted from major disasters. Prior to the act, congressional action after each individual incident was needed to provide federal aid. Once in place, the law authorized the President to make the decision to provide federal aid without the consent of Congress. The act also put in place the standard process in which the governors could ask the President for federal disaster assistance for their respective states.[17]

### *Disaster Relief Act of 1966 (P.L. 89-769)*

To some, the Disaster Act of 1950 appeared to effectively handle routine disasters, but did not adequately address large-scale, catastrophic disasters. In response to this need, Congress enacted the Disaster Relief Act of 1966. Some of the measures in the act included the authorization of federal agencies to provide disaster loans below market rates, and the extension of aid to unincorporated areas.

### *Disaster Relief Act of 1974 (P.L. 93-288)*

The Disaster Relief Act of 1974 contained several precedent-setting features. The act created the first program to provide direct assistance to individuals and households following a disaster. It instituted the Individual and Family Grant (IFG) program, which supplied 75% of the funding for state-administered programs that provided money to purchase clothing, furniture, and essential needs following a disaster. The act also formalized efforts to mitigate disasters, as opposed to merely responding to them. Additionally, the act stressed a multi-hazard approach wherein government officials would prepare and respond to all types of disasters, rather than maintaining separate capacities for different types of hazards.[18]

### *The Robert T. Stafford Disaster Relief and Emergency Assistance Act (P.L. 93-288)*

Today, the Robert T. Stafford Disaster Relief and Emergency Assistance Act constitutes the statutory authority for most federal disaster response and recovery activities especially as they pertain to FEMA and FEMA programs. The act authorizes the President to issue major disaster, emergency, and fire management declarations at the request of the states and tribal governments.[19] The declarations enable federal agencies to provide assistance to states overwhelmed by disasters. Stafford Act disaster assistance is provided through funds appropriated to the DRF. The funds may be used by states, localities, and certain non-profit organizations for activities such as providing mass care, restoring damaged or destroyed facilities, clearing debris, aiding individuals and families with uninsured needs, and mitigating the impact of future disasters.[20]

The history of federal disaster relief legislation demonstrates a steady expansion of federal aid since the 1950s. As the federal role in emergencies and major disasters expanded, so too have the costs. The costs associated with disaster relief have led some to contemplate changing how the federal government funds these events. Some of these proposals are discussed later in this report.

## *Post-Katrina Emergency Management Reform Act of 2006 (P.L. 109-295)*

In the aftermath of Hurricane Katrina, Congress passed the Post-Katrina Emergency Management Reform Act[21] to address emergency preparedness and response shortcomings identified in the reports published by congressional committees and the White House. Based on those reports and oversight hearings on many aspects of FEMA's performance during the hurricane season of 2005, Congress expanded the authority of the FEMA administrator, authorized accelerated federal aid, and raised some ceilings on federal assistance, among other changes. In addition, Congress mandated that, for FY2007, FEMA had to submit to Congress a monthly report on the DRF detailing allocations, obligations, and expenditures for Hurricanes Katrina, Rita, and Wilma.[22] The reports had to include information on national flood insurance claims, manufactured housing data, obligations, allocations for housing assistance, public assistance, individual assistance and expenditures by state for unemployment.[23]

## HOW THE DRF IS FUNDED

FEMA receives appropriations for disaster relief through annual appropriations. Each fiscal year, FEMA and the Office of Management and Budget (OMB) submit a request to the President for the amount of funding the two agencies determine the DRF should receive. The President then submits an Administration request to Congress. The President may, or may not, use the amount suggested by FEMA and the OMB.

The methodology previously used to determine the budget recommendation incorporated four data points. The data points included (1) the available appropriation, (2) the DRF monthly average (the amount in the

DRF), (3) the monthly cost estimates for catastrophic events, and (4) the estimated monthly recoveries of unobligated funds.[24]

1. **Available Appropriation.** The available appropriation was a combination of prior-year funds that are carried over, the current fiscal year appropriation, and any supplemental appropriation funding.[25]
2. **DRF Monthly Average.** The calculation for the DRF monthly average was based on a five-year rolling average "normal" disaster's costs. Normal disasters were considered to be incidents that cost less than $500 million. The rationale of excluding large events from the calculation is discussed later in this report.
3. **Monthly Cost Estimates for Catastrophic Events.** Estimates obtained from the field on pending (still open) disaster projects were routinely used in calculating monthly cost estimates.
4. **Estimated Recoveries.** Estimated recoveries represent the recovery of obligated funds that have not been used. This could include duplication of benefit funds[26] as well as long-term projects for PA or mitigation that either were not finished, or were completed at a lower cost.

The end-of-fiscal-year projection was estimated by subtracting the cumulative DRF monthly averages and cost estimates for incidents from the available appropriation. Then, the cumulative recoveries were added. The DRF end-of-fiscal-year estimate was then revised on a monthly basis taking into consideration the actual obligations that were recorded in lieu of the monthly estimates, and new estimates submitted for "open" incidents.[27]

The methodology described above may have led to a series of budget requests that were too low to meet disaster assistance needs solely through regular appropriations. In particular, the elimination of incidents of $500 million or more from the calculation may have led to a lower budget projection. The passage of the Budget Control Act (P.L. 112-25, hereinafter the BCA) may have altered how the DRF is now being budgeted. The BCA includes a series of provisions that directed the Office of Management and Budget (OMB) to annually calculate an "allowable adjustment" for disaster relief to the BCA's discretionary spending caps. That adjustment, if used, would make additional budget authority available for the federal costs incurred by major disasters declared under the Stafford Act beyond what is allowed in the regular discretionary budget allocation. The OMB calculation may have provided a mechanism that encourages a larger regular appropriation to the DRF. It is possible that larger DRF appropriations may reduce the need for supplemental appropriations. In addition, in the past there may have been more pressure in previous years to quickly pass a supplemental appropriation to meet the needs of the disaster. Higher appropriations for the DRF through regular appropriations may provide Congress with more time to debate disaster needs and target disaster assistance needs more efficiently when large scale disasters occur. For example, when Hurricane Sandy struck the northeastern United States in October 2012, the DRF had roughly $7 billion to meet the immediate demands of the hurricane. In previous years, when a large scale disaster occurred, the DRF balance was generally low due to smaller regular appropriations to the account.

A comparison between Administration requests and annual appropriations is presented in **Table 1**.

## THE DEBATE OVER EMERGENCY SUPPLEMENTAL APPROPRIATIONS

Congress has traditionally appropriated funds to maintain the DRF at a certain level, and then provided additional financing for assistance through supplemental appropriations following a specific large disaster. Currently, the DRF funds disaster relief for emergencies and major disasters that cost $500 million or less. Major disasters costing more than $500 million are generally funded with emergency supplemental appropriations. A supplemental appropriation provides additional budget authority during the current fiscal year either to finance activities not provided for in the regular appropriation, or to provide funds when the regular appropriation is deemed insufficient.

Catastrophic incidents may deplete the DRF. In such cases, the President may submit a request to Congress for an emergency supplemental appropriation. Historically, FEMA is the second-largest recipient of supplemental appropriations.[28] Most emergency supplemental appropriations originate with the Administration. Some emergency supplemental appropriations, however, have been initiated by Congress. For example, in 2007, Congress waived the cost share provisions for Hurricanes Katrina and Rita, which prompted a supplemental appropriation (P.L. 110-28). Another example of an emergency supplemental appropriation originating in Congress was in 2008, when Congress added DRF supplemental funding (P.L. 110-329) to the FY2009 Department of Homeland Security Appropriations bill on its own initiative in response to the Midwest Flooding and Hurricanes Ike and Gustav. The amount of funding for the supplemental appropriation was based on estimates provided to Congress by FEMA.

The current process of funding disasters has been a subject of debate at times because of concern over high levels of federal spending and the inability to institute fiscal planning mechanisms. Some of the debate includes discussions of options for lowering the budget deficit through budgetary enforcement procedures in Congress. Others argue supplemental appropriations are used to make budgets appear to forecast a lower deficit. OMB indicated a desire to break from the current practice of funding disaster relief by budgeting for large-scale disasters. *In A New Era of Responsibility*, the first FY2010 document issued by the Obama Administration, OMB stated that "in the past, budgets assumed that there would not be any natural disasters in our nation that would necessitate federal help.... This omission is irresponsible, and has permitted past Administrations to project deficits that were lower than likely occur."[29]

Accordingly, the debate over the amount of funding in the DRF generally falls into two basic categories: (1) proponents who advocate the use of emergency supplemental appropriations to augment the DRF, and (2) those who oppose their use.

Proponents of the current system of using emergency supplemental appropriations argue there are several benefits associated with their use. Some of these perceived benefits are summarized as follows:

- Disasters cannot be anticipated, a condition that creates a budgeting challenge. Because of their flexibility, emergency supplemental appropriations facilitate funding decisions for disasters, the timing and severity of which are unpredictable. Budgeting for large disasters through regular appropriations would likely require Congress to reduce funding for other programs to pay for an unknown and possibly non-existent future event.
- The President is authorized to unilaterally determine when federal assistance ismade available after an incident. Congress retains authority to control spendingby voting on supplemental funding legislation. In essence, a balance of powers is at least theoretically maintained through this process.
- A large balance in the DRF may be subject to transfer or rescission.[30] Such actions could carry an additional negative consequence if a large

disaster were to take place after funds have been withdrawn. An emergency supplemental can besized according to the needs of the actual incident.

Arguments used by some of the opponents against the enactment of emergency supplemental appropriations to augment the DRF include the following:

- Supplemental spending allows lawmakers to circumvent budgetary enforcement mechanisms by deliberately underfunding programs in annual appropriations. Because there is a perception that a supplemental appropriation must be passed quickly, supplemental appropriations may be less likely to be scrutinized in the same manner as an annual appropriation.[31]
- Reliance upon emergency supplemental appropriations may result in unnecessarily high funding levels, as early damage estimates may overstate actual needs. Maintaining DRF funds at a level that more accurately reflects the needs would foster fiscal responsibility because all disaster response would have to be budgeted with existing funds.
- Use of emergency supplemental appropriations may give the appearance of fiscal irresponsibility on the part of the federal government, especially when they are frequent and large. Peter Orszag, former Director of the Office of Management and Budget (OMB), voiced this concern when he stated: "The first step in addressing this very deep fiscal hole is honesty. This budget will not play the games that are typically played in which you assume that there will never again be a hurricane or disaster."[32]
- Emergency supplemental appropriations provide a vehicle for non-germane provisions that may not pass on their own, or make the appropriation contentious, thus slowing down federal recovery assistance after a disaster.

Comparative data on Administration requests, annual appropriations, and supplemental appropriations for the DRF are provided in **Table 1** of this report.

## Table 1. Disaster Relief Fund Requests, Appropriations, and Supplemental Appropriations (in millions of dollars)

| Fiscal Year | Administrative Request (Nominal) | Administrative Request (2012 Constant Dollars) | Enacted Appropriation (Nominal) | Enacted Appropriation (2012 Constant Dollars) | Supplemental (Nominal) | Supplemental (2012 Constant Dollars) | Total Nominal | Total (Constant 2012 Dollars) |
|---|---|---|---|---|---|---|---|---|
| 1989 | 200 | 333 | 100 | 166 | 1,108 | 1,845 | 1,208 | 2,010 |
| 1990 | 270 | 434 | 98 | 157 | 1,150 | 1,847 | 1,248 | 2,003 |
| 1991 | 270 | 418 | 0 | 0 | 0 | 0 | 0 | 0 |
| 1992 | 184 | 277 | 185 | 279 | 3,993 | 6,021 | 4,178 | 6,296 |
| 1993 | 292 | 431 | 292 | 431 | 1,735 | 2,560 | 2,027 | 2,989 |
| 1994 | 1,154 | 1,667 | 292 | 422 | 5,117 | 7,393 | 5,409 | 7,810 |
| 1995 | 320 | 453 | 318 | 450 | 2,275 | 3,219 | 2,593 | 3,666 |
| 1996 | 320 | 444 | 222 | 308 | 3,171 | 4,402 | 3,393 | 4,707 |
| 1997 | 320 | 436 | 1,320 | 1,799 | 3,300 | 4,498 | 4,620 | 6,293 |
| 1998 | 2,708 | 3,645 | 320 | 431 | 1,600 | 2,153 | 1,920 | 2,582 |
| 1999 | 2,566 | 3,408 | 308 | 409 | 1,806 | 2,399 | 2,114 | 2,806 |
| 2000 | 2,780 | 3,621 | 2,780 | 3,621 | 0 | 0 | 2,780 | 3,618 |
| 2001 | 2,909 | 3,702 | 1,594 | 2,028 | 0 | 0 | 1,594 | 2,027 |
| 2002 | 1,369 | 1,714 | 664 | 831 | 0 | 0 | 664 | 831 |
| 2003 | 1,843 | 2,261 | 800 | 981 | 1,426 | 1,749 | 2,226 | 2,729 |
| 2004 | 1,956 | 2,340 | 1,800 | 2,154 | 2,213 | 2,648 | 4,013 | 4,798 |
| 2005 | 2,151 | 2,493 | 2,042 | 2,366 | 43,091 | 49,934 | 45,133 | 52,264 |
| 2006 | 2,140 | 2,398 | 1,770 | 1,984 | 6,000 | 6,724 | 7,770 | 8,702 |
| 2007 | 1,941 | 2,113 | 1,500 | 1,633 | 4,092 | 4,454 | 5,592 | 6,083 |
| 2008 | 1,652 | 1,757 | 1,400 | 1,489 | 10,960 | 11,659 | 12,360 | 13,140 |
| 2009 | 1,900 | 1,996 | 1,278 | 1,342 | 0 | 0 | 1,299 | 1,363 |
| 2010 | 2,000 | 2,079 | 1,600 | 1,664 | 5,100 | 5,303 | 6,700 | 6,962 |
| 2011 | 1,950 | 1,986 | 2,650 | 2,699 | 0 | 0 | 2,650 | 2,697 |
| 2012 | 1,800 | 1,800 | 700 | 700 | 6,400 | 6,400 | 7,100 | 7,095 |
| 2013 | 6,089 | 6,089 | 7,007 | 7,007 | 11,485 | 11,485 | 17,574 | 17,574 |
| Total | 41,084 | 48,295 | 31,040 | 35,351 | 116,022 | 136,693 | 146,165 | 171,045 |

Source: CRS analysis of Administration budget documents and appropriations statutes.

Note: FY2012 current dollar conversions were calculated using GDP Chained Price Index data in Table 10.1, FY2014 Budget Historical Tables volume available at http://www.whitehouse.gov/omb/budget/Historicals.

# EMERGENCY SUPPLEMENTAL APPROPRIATIONS: FY1989-FY2013

This section provides additional summary information on emergency supplemental appropriations legislation enacted since 1989; it uses a broad concept of what constitutes emergency disaster assistance. The funds cited include appropriations to all federal agencies that undertook disaster relief, repair of federal facilities, and hazard mitigation activities directed at reducing the impact of future disasters. Funds used for activities such as research or administrative costs have been omitted from this analysis in an attempt to focus solely on disaster relief and assistance. Moreover, counterterrorism, law enforcement, and national security appropriations are not included in this compilation. Unless otherwise noted, this report does not take into account rescissions approved by Congress after funds have been appropriated for disaster assistance.

Since FY1989, Congress has appropriated roughly $309 billion for disaster assistance in 37 appropriations acts, primarily supplemental appropriations acts after significant catastrophes occurred in the United States. The mean annual appropriation is approximately $12.9 billion. The mean supplemental appropriation for all 37 bills is approximately $8.3 billion. Current dollar figures for each appropriation and totals for the period appear in **Table 2**.

As reflected in **Table 2**, supplemental appropriations have generally been enacted as stand-alone legislation. In some instances, however, emergency disaster relief funding has been enacted as part of regular appropriations measures, continuing appropriations acts (continuing resolutions), or as a part of omnibus appropriations legislation. Requested funding levels noted in the third column of **Table 2** reflect House Appropriations Committee data on total requested funding for the entire enacted bill.

## Recent Enacted Supplemental Appropriations

In the 113th Congress, H.R. 152, the Disaster Relief Appropriations Act, 2013, was introduced in the House by Representative Harold Rogers on January 4, 2013. The bill provided disaster assistance through numerous federal agencies and entities in response to Hurricane Sandy. The bill included provided $11.5 billion for the DRF, $6 million for the Commodity Assistance

Program account—specifically for The Emergency Food Assistance Program (TEFAP), and $13.07 billion for the Department of Transportation. The bill was enacted as P.L. 113-2.[33]

**Table 2. Emergency Appropriations for Disaster Relief: All Agencies FY1989 – Present (Dollars in Thousands)**

| Fiscal Year | Disaster Event and Date of Major Disaster Declaration | Date Signed into Law and P.L. Number | Emergency Assistance Funding | Emergency Assistance Funding (2012 dollars) |
|---|---|---|---|---|
| 2013 | Hurricane Sandy, Nov. - Dec. 2012 | Jan. 29, 2013, P.L. 113-2 | $50,600,000 | $50,600,000 |
| 2012 | Storms, Flooding, Drought, and Hurricane Irene events in 2011 | Dec. 23, 2011, P.L. 112-77 | $8,124,000 | $8,124,000 |
| 2010 | Hurricane Katrina, severe storms/ flooding, wildfires, oil spill, various dates | July 19, 2010 P.L. 111-212 | $5,563,600 | $5,784,746 |
| 2008 | Hurricane Katrina, Midwest Flooding and the 2008 hurricanes, various dates | Sept. 30, 2008 P.L. 110-329 | $23,389,800 | $24,882,126 |
| 2008 | Hurricane Katrina and other hurricanes in the 2005 season | June 30, 2008 P.L. 110-252 | $8,381,805 | $8,916,585 |
| 2008 | Hurricane Katrina & California Wildfires, Oct. 24, 2007 | Nov. 13, 2007 P.L. 110-116 | $6,355,000 | $6,739,188 |
| 2007 | Hurricane Katrina, Aug. 29, 2005 | May 25, 2007 P.L. 110-28 | $7,679,000 | $8,358,468 |
| 2006 | Hurricanes Katrina, Rita, Wilma, Aug. - Sept. 2005 June 15, 2006 P.L. 109-234 | $19,340,000 | $21,674,267 | |

| Fiscal Year | Disaster Event and Date of Major Disaster Declaration | Date Signed into Law and P.L. Number | Emergency Assistance Funding | Emergency Assistance Funding (2012 dollars) |
|---|---|---|---|---|
| 2006 | Hurricanes Katrina, Rita, Wilma, Aug. - Sept. 2005 | Dec. 30, 2005 P.L. 109-148 | $29,046,985 | $32,552,849 |
| 2005 | Hurricane Katrina Aug. 29, 2005 | Sept. 8, 2005 P.L. 109-62 | $51,800,000 | $60,025,840 |
| 2005 | Hurricane Katrina Aug. 29, 2005 | Sept. 2, 2005 P.L. 109-61 | $10,500,000 | $12,167,400 |
| 2005 | Hurricanes Ivan, Jeanne, Sept. 1, 2004 | Oct. 13, 2004 P.L. 108-324 | $11,103,887 | $12,867,184 |
| 2004 | Hurricanes Charley, Frances, Sept. 1, 2004 | Sept. 8, 2004 P.L. 108-303 | $2,000,000 | $2,392,979 |
| 2004 | Wildfires, various dates | Aug. 8, 2004 P.L. 108-287 | $500,000 | $598,245 |
| 2004 | Hurricane Isabel Sept. 18, 2003 | Nov. 6, 2003 P.L. 108-106 | $813,000 | $972,746 |
| 2003 | Storms, various 2003 dates | Sept. 30, 2003 P.L. 108-83 | $820,700 | $1,006,804 |
| 2003 | Tornadoes, May 6, 2003 | Aug. 8, 2003 P.L. 108-69 | $983,600 | $1,206,644 |
| 2002 | Terrorist attacks, Sept. 11, 2001 | Aug. 2, 2002 P.L. 107-206 | $6,167,600 | $7,720,660 |
| 2001 | Terrorist attacks, Sept. 11, 2001 | Sept. 18, 2001 P.L. 107-38 | $20,000,000 | $25,451,351 |
| 2001 | Nisqually Earthquake | July 24, 2001 P.L. 107-20 | $365,700 | $465,378 |
| 2000 | Hurricane Floyd Sept. 16, 1999 | Oct. 20, 1999 P.L. 106-74 | $2,480,425 | $3,230,658 |
| 1999 | Tornadoes, various dates | May 21, 1999 P.L. 106-31 | $1,296,723 | $1,722,424 |
| 1999 | Hurricanes Georges, Bonnie, flooding, various dates | Oct. 21, 1998 P.L. 105-277 | $1,830,977 | $2,432,068 |

## Table 2. (Continued)

| Fiscal Year | Disaster Event and Date of Major Disaster Declaration | Date Signed into Law and P.L. Number | Emergency Assistance Funding | Emergency Assistance Funding (2012 dollars) |
|---|---|---|---|---|
| 1998 | El Niño floods, Feb. 9, 1998 | May 1, 1998 P.L. 105-174 | $2,602,173 | $3,502,204 |
| 1997 | Dakotas flooding, Apr. 7, 1997 | June 12, 1997 P.L. 105-18 | $5,863,883 | $7,992,317 |
| 1995 | Oklahoma City bombing, Apr. 25, 1995 July 27, 1995 P.L. 104-19 | | $6,599,531 | $9,337,651 |
| 1995 | Northridge Earthquake, Tropical Storm Alberto, various dates | Sept. 28, 1994 P.L. 103-327 | $417,500 | $590,719 |
| 1994 | Midwest floods, CA fires, and Northridge earthquake Jan. 17, 1994 | Feb. 12, 1994 P.L. 103-211 | $8,837,952 | $12,769,849 |
| 1993 | Midwest floods, June 11, 1993 | Aug. 12, 1993 P.L. 103-75 | $3,494,750 | $5,156,247 |
| 1993 | Hurricanes Andrew, Iniki, various dates | July 2, 1993 P.L. 103-50 | $52,345 | $78,930 |
| 1992 | Hurricanes Andrew, Iniki, Aug. 24, 1992 | Sept. 23, 1992 P.L. 102-368 | $5,767,116 | $8,696,075 |
| 1992 | L.A. riots/Chicago flood, various dates | June 22, 1992 P.L. 102-302 | $469,650 | $708,172 |
| 1992 | Hurricane Bob, various dates | Dec. 12, 1991 P.L. 102-229 | $943,000 | $1,421,924 |
| 1990 | Hurricane Hugo, Exxon Valdez, various dates | May 25, 1990 P.L. 101-302 | $670,412 | $1,076,598 |

| Fiscal Year | Disaster Event and Date of Major Disaster Declaration | Date Signed into Law and P.L. Number | Emergency Assistance Funding | Emergency Assistance Funding (2012 dollars) |
|---|---|---|---|---|
| 1990 | Hurricane Hugo, Loma Prieta Earthquake, Oct. 18, 1989 | Oct. 26, 1989 P.L. 101-130 | $2,850,000 | $4,576,746 |
| 1989 | Hurricane Hugo, Sept. 20, 1989 | Sept. 29, 1989 P.L. 101-100 | $1,108,000 | $1,779,310 |
| 1989 | Fires on federal lands, various dates | June 30, 1989 P.L. 101-45 | $348,969 | $581,013 |
| **Total** | | | $309,168,083 | $358,160,365 |

Source: Supplemental funding totals derived, in part, from CRS analysis of emergency appropriations after disasters, FY1989-FY2013.

a. Notes: Declaration dates in this table represent the date the President issued a major disaster declaration for the disaster that appeared to be the primary catalyst for the supplemental appropriations legislation. In a series of disasters (such as the Midwest floods of 1993) this date represents the first of several declarations associated with that particular disaster. In some instances, identifying which disasters were primarily associated with consideration of the supplemental appropriations was not possible. FY2012 current dollar conversions were calculated using GDP Chained Price Index data in Table 10.1, FY2014 Budget Historical Tables volume available at http://www.whitehouse.gov/omb/budget/Historicals.

In the 112th Congress H.R. 3672, the Disaster Relief Appropriations Act, 2012, was introduced in the House by Representative Harold Rogers on December 14, 2001. The bill included $6.4 billion for the DRF and roughly $1.7 billion for the Army Corps of Engineers to repair damages to federal projects resulting from major disasters, operations and expenses, and other projects to prepare for floods, hurricanes, and other natural disasters. The bill was enacted as P.L. 112-77 on December 23, 2011.

In the 111th Congress, H.R. 4899, the Disaster Relief and Summer Jobs Act of 2010, was introduced in the House by Representative David Obey on March 21, 2010.

The bill included $5.1 billion for the DRF. During consideration H.R. 4899 became a vehicle for additional non-disaster funding including $33 billion in war funding for the Department of Defense, and funding for court case relief for veterans, Native Americans, and minority farmers. The bill was signed into law on July 29, 2010, and became P.L. 111-212.

In addition to the $5.1 billion provided for the DRF, other types of disaster assistance in P.L. 111-212 included $18 million for forest restoration, $49 million for the Economic Development Administration (EDA) for long-term recovery and infrastructure restoration for floods, and another $5 million for technical assistance for activities associated with the Deepwater Horizon oil spill.

The act provided $100 million to the Department of Housing and Urban Development (HUD) for housing and economic revitalization, and roughly $26 million was provided to the Department of the Interior to address environmental and fishery impacts resulting from the Deepwater Horizon oil spill.

In the 110$^{th}$ Congress, President George W. Bush signed into law four measures (P.L. 110-28, P.L. 110-116, P.L. 110-252, and P.L. 110-329) that provided roughly $45 billion in supplemental appropriations for disaster relief and recovery (most of it for the DRF). P.L. 110-28, signed on May 25, 2007, included an appropriation of $7.6 billion for disaster assistance, $6.9 billion of which was classified for Hurricane Katrina recovery. P.L. 110-116, signed into law on November 13, 2007, provided a total of $6.3 billion for continued recovery efforts related to Hurricanes Katrina, Rita, and Wilma, and for other declared major disasters or emergencies. P.L. 110-116 also included $500 million for firefighting expenses related to the 2007 California wildfires. P.L. 110-252, signed into law June 30, 2008, provided $8.4 billion in disaster assistance, most of which was directed at continuing recovery needs resulting from the 2005 hurricane season.

P.L. 110-329, signed into law on September 30, 2008, included an appropriation for emergency and disaster relief of $23.4 billion. Of this amount, roughly $2.9 billion was continued disaster relief for the 2005 hurricane season. The largest share of the funding (just over $8.8 billion), however, was for a string of disasters that occurred in 2008 including Hurricanes Gustav and Ike, wildfires in California, and the Midwest floods. One of the largest funding components in P.L. 110-329 was designated for the Department of Housing and Urban Development's (HUD's) Community Development Fund, which received $6.5 billion specifically for disaster relief, long-term recovery, and economic revitalization for areas affected by the 2008 disasters.

Other funding in the law included $135 million for wildfire suppression, and a $100 million direct appropriation for the American Red Cross for reimbursement of disaster relief and recovery expenditures associated with emergencies and disasters that took place in 2008.[34]

## Supplemental Appropriations for Hurricanes Katrina, Rita, and Wilma

Since FY2005, a number of supplemental appropriations have been enacted to provide the Gulf Coast disaster relief and recovery assistance after Hurricanes Katrina, Rita, and Wilma. The following section provides examples of federal assistance provided in each supplemental appropriation.[35]

### *Measures Enacted in FY2010*

P.L. 111-212 provided FEMA funding for ongoing recovery projects in the Gulf Coast and to pay claims awarded by arbitrators to state, local, and nonprofits for damages resulting from Hurricane Katrina. The claims were the result of damage disputes. In the case of Louisiana's Charity Hospital, FEMA questioned whether all the damages to the aging hospital were storm-related and offered the state $150 million for repairs. The state disagreed and appealed the decision. The arbiters agreed with the state and ruled that FEMA must pay the state $475 million for repairs to the hospital.

Although some of the disaster assistance in P.L. 111-212 funds Gulf Coast recovery, the amount of assistance for projects related to the 2005 hurricane season is unclear. Thus, funding data from P.L. 111-212 have been excluded from the following summary section and **Table 3**.

### *Measures Enacted in FY2008*

On June 30, 2008, Congress enacted the Supplemental Appropriations Act, 2008 (P.L. 110-252). Some of the funding in P.L. 110-252 includes $100 million for the Economic Development Administration's economic development assistance programs, $73 million for the Louisiana Road Home Program, and $300 million for HUD's Community Development fund. The majority of disaster assistance funding (over $6 billion) in P.L. 110-252 was directed to the Corps of Engineers for projects aimed at repairing damages incurred during the 2005 hurricane season, as well as programs designed to mitigate against future hurricanes.[36]

Another supplemental, the Consolidated Security, Disaster Assistance, and Continuing Appropriations Act of 2009, was passed three months later on September 30, 2008 (P.L. 110-329). P.L. 110-329 includes ongoing disaster relief for destruction resulting from the 2005 hurricane season, including $85 million for the Tenant-Based Assistance Program and $15 million for the Public Housing Capital Fund (administered by HUD), and $15 million for school education programs to help local educational agencies with increased homeless students enrollments as a result of the 2005 hurricanes (administered by the Department of Education).

The amount provided in the statute for disaster relief as a result of the 2005 hurricane season is roughly $2.9 billion.[37]

***Measures Enacted in FY2007***

On May 25, 2007, the President signed into law P.L. 110-28, which appropriated $120 billion in emergency supplemental funding for Iraq, Afghanistan, and other matters, including $6.9 billion for continued Gulf Coast relief and waived cost-shares for the Gulf Coast states. The measure was a successor to previous emergency supplemental legislation in the 110th Congress, H.R. 1591, which had been vetoed by the President on May 1, 2007. This was the fifth supplemental measure enacted in the 110th Congress containing disaster assistance specifically provided in response to Hurricanes Katrina and Rita.

The sixth supplemental measure enacted as part of P.L. 110-116 on November 13, 2007, provided an additional $6.3 billion for emergency assistance, most, but not all, of which can be attributed to the Gulf Coast recovery. The $3 billion appropriated for HUD's Community Planning and Development Fund can only be used for the Louisiana Road Home program,[38] but the $2.9 billion appropriated for the DRF can be used not only for the Gulf Coast but for other declared disasters as well.

After the enactment of P.L. 110-252, the total amount appropriated by Congress in supplemental funding after the 2005 hurricanes surpassed the $130 billion mark. In addition to these rescissions and appropriations, Congress enacted other funding changes by transferring $712 million from FEMA to the Small Business Administration for disaster loans (P.L. 109-174).

***Measures Enacted in FY2006***

Later, when Hurricanes Rita and Wilma struck, Congress enacted two additional emergency supplemental appropriations; the costs of both were offset by rescissions.

The FY2006 appropriations legislation for the Department of Defense (P.L. 109-148) rescinded roughly $34 billion in funds previously appropriated (almost 70% of which was taken from funds previously appropriated to DHS) and appropriated $29 billion to other accounts primarily to pay for the restoration of federal facilities damaged by the hurricanes).[39] Also in FY2006, Congress agreed to an Administration request for further funding—$19.3 billion was appropriated in supplemental legislation (P.L. 109-234) for recovery assistance, with roughly $64 million rescinded from two accounts ($15 million from flood control, Corps of Engineers, and $49.5 million from Navy Reserve construction, Department of Defense).

*Measures Enacted in FY2005*

In response to the widespread destruction caused by three catastrophic hurricanes at the end of the summer of 2005, Congress enacted four emergency supplemental appropriations bills. Two of the bills were enacted as FY2005 emergency supplemental appropriations after Hurricane Katrina devastated parts of Florida and Alabama and resulted in presidential major disaster declarations for all jurisdictions in Louisiana and Mississippi. The two emergency supplemental appropriations (P.L. 109-61 and P.L. 109-62) together provided $62.6 billion for emergency response and recovery needs; most of the appropriations in these two bills funded the DRF.

## Emergency Supplemental Appropriations for Hurricanes Katrina, Rita, and Wilma by Federal Agency

The primary recipient of emergency supplemental appropriations related to Hurricanes Katrina, Rita, and Wilma is DHS, which has received 57% of emergency supplemental funding for disaster assistance. HUD received 16%, DOD Army Corps of Engineers received 13%, the DOD received another 7% for miscellaneous activities, and the Department of Transportation received 3%.[40] **Table 3** provides information on the appropriations made in the eight emergency supplemental appropriations enacted largely to address losses associated with Hurricanes Katrina, Rita, and Wilma and identify the departments and agencies that received funding for disaster relief.[41]

## Table 3. FY2005-FY2008 Supplemental Disaster Appropriations after Hurricanes Katrina, Rita, and Wilma
### (thousands of current dollars)

| Department | FY2005 P.L. 109-61 | FY2005 P.L. 109-62 | FY2006 P.L. 109-148 | FY2006 P.L. 109-234 | FY2007 P.L. 110-28 | FY2007 P.L. 110-116 | FY2008 P.L. 110-252 | FY2008 P.L. 110-329 | TOTAL Department |
|---|---|---|---|---|---|---|---|---|---|
| Agriculture | | | $1,183,000 | $152,000[a] | | | | $38,000 | $1,373,000 |
| Department of Commerce | | | $55,000 | $150,000 | $110,000 | | $100,000 | $481,000 | $896,000 |
| Defense-Military | $500,000 | $1,400,000 | $5,754,000 | $1,488,000[b] | | | | | $9,142,000 |
| Defense-Civil/ Corps of Engineers | | $400,000 | $2,900,000 | $3,686,000[c] | $1,433,000 | | $6,366,988 | $1,621,200 | $16,407,188 |
| Education and related agencies | | | $1,600,000 | $285,000 | $60,000 | | | $30,000 | $1,975,000 |
| Health and Human Services | | | $640,000 | $12,000 | | | | $600,000[d] | $1,252,000 |
| Homeland Security | $10,000,000 | $50,000,000 | $285,000 | $6,662,000 | $4,110,000 | $2,900,000 | | $20,000 | $73,977,000 |
| Housing and Urban Development | | | $11,890,000 | $5,200,000 | $7,000 | $3,000,000 | $373,000 | $150,000 | $20,620,000 |
| Interior | | | $70,000 | $256,000 | $10,000 | [c] | | | $336,000 |
| Justice | | | $229,000 | $9,000 | $50,000 | | | | $288,000 |
| Labor | | | $125,000 | $16,000 | | | | | $141,000 |
| Transportation | | | $2,798,000 | $702,000[f] | $906,020 | | | | $4,406,020 |
| Veterans Affairs | | | $658,000 | $586,000 | $14,500 | | | | $1,258,500 |
| Armed Forces Retirement Home | | | | $176,000 | | | | | $176,000 |
| Corp. for National & Community Svc | | | | $10,000 | | | | | $10,000 |
| Environmental Protection Agency | | | $8,000 | $13,000 | | | | | $21,000 |

| Department | FY2005 P.L. 109-61 | FY2005 P.L. 109-62 | FY2006 P.L. 109-148 | FY2006 P.L. 109-234 | FY2007 P.L. 110-28 | FY2007 P.L. 110-116 | FY2008 P.L. 110-252 | FY2008 P.L. 110-329 | TOTAL Department |
|---|---|---|---|---|---|---|---|---|---|
| General Services Administration | | | $38,000 | $37,000 | | | | | $75,000 |
| Historically Black Colleges Cap. Financing | | | | $15,000 | | | | | $15,000 |
| Nat'l Aeronautics & Space Admin. | | | $350,000 | $35,000 | $20,000 | | | | $405,000 |
| Judiciary | | | $18,000 | | | | | | $18,000 |
| Small Business Administration | | | $446,000 | $542,000 | $181,070 | | | | $1,169,070 |
| Total | $10,500,000 | $51,800,000 | $29,047,000 | $20,032,000 | $6,901,590 | $5,900,000 | $6,839,989 | $2,910,200 | $133,960,778 |

Source: Congressional Research Service. Numbers have been rounded.

a. Does not include authority for $500 million in direct assistance to be drawn from the Commodity Credit Corporation, authorized in Title III of P.L. 109-234.
b. Includes rescissions and military construction accounts.
c. Includes rescissions.
d. This funding amount is divided between the 2005 hurricane season, and hurricanes, floods, and other disasters that occurred in 2008.
e. In Division B of P.L. 110-116, 121 Stat. 1342-1343, Section 157 provides $329 million for Forest Service Wildland Fire Management and $171 million for Bureau of Land Management Wildland Fire Management. This funding is not included in the table since the funding was for wildland firefighting activities and not related to Gulf Coast hurricane relief and recovery.
f. Department of Transportation funds derived from Highway Trust Fund rescission.

## ISSUES FOR CONGRESS

### Authorizing the DRF

As mentioned earlier in this report, the DRF has not been authorized explicitly by Congress. Authorization measures are generally used to establish, continue, or modify an agency or program for a fixed or an indefinite period of time. The measures are also used to explicitly name accounts or programs.[42] Authorizations also establish the duties and functions of an agency or program, its organizational structure, and the responsibilities of agency or program officials. One function of such legislation is to authorize subsequent spending in appropriations bills for specific agencies, programs, projects, accounts, or funds; such authorizations may include spending ceilings.[43]

It may be argued that a large account such as the DRF should be authorized specifically by name to ensure these functions are formally established. Others may view this as unnecessary because they believe the Stafford Act, which authorizes appropriations to carry out disaster relief, preparedness, and hazard mitigation activities, is sufficient.[44]

If authorizing the DRF became an issue of active debate, Congress could consider legislation similar to Section 329 of H.R. 3377 (introduced in the 111th Congress) to authorize the DRF, or it could elect to maintain the DRF as described under 42 U.S.C. 5197e of the Stafford Act.

### Restructuring Budgetary Procedures

Some proposals have been advanced to reduce the need for emergency supplemental appropriations. For example, the Bipartisan Task Force on Funding Disaster Relief[45] offered several policy options for changing budgetary procedures for funding disaster assistance. Some of these options for restructuring the budgetary process include the following:

- **Eliminate any Nonemergency Funding in a Bill.** Many emergency supplemental bills include a variety of funding and other needs in a single bill. Eliminating nonemergency (or nongermane) elements from legislation could prevent the passage of legislation that might not pass Congress if it were not attached to an emergency supplemental appropriation. Such action may make emergency

supplemental appropriations less controversial and ensure expedited enactment.
- **Strengthening Criteria.** The Bipartisan Task Force offered the use of more strict criteria as a policy option. If specific criteria were enforced, it is argued, Congress and the President would have to issue written justifications for designating appropriations as emergencies. Proponents of rewriting the criteria say the measure would open the debate on the use of emergency supplemental appropriations by adding an additional layer of scrutiny to the emergency appropriation process. Opponents may argue that changing the criteria may create unnecessary delays for appropriating needed emergency funding.
- **Increase Funding to the DRF.** As mentioned earlier in this report, Congress may decide to increase the funding level of the DRF through annual appropriations. Doing so could eliminate, or at least greatly decrease, the need for emergency supplemental funding.
- **The Creation of a Rainy-Day Fund.** A rainy-day fund, also known as a reserve account, would be financed by cuts in other discretionary accounts, or through revenue raising measures. Spending from this fund would then be allowed only when needed for expensive disasters. Proponents of this policy option would likely argue a rainy-day fund carries several advantages. For one, in contrast to emergency supplemental appropriations which increase the federal deficit through borrowing funds, rainy-day accounts do not add to the federal deficit because they are funded through savings and/or revenue raising measures. Furthermore, the balance for a rainy-day fund would increase during periods in which there are few, or relatively small disasters. Another advantage would be that a rainy-day fund would not require a restructuring of the process of administering disaster relief.
- **The Creation of a Contingency Fund.** A contingency fund based on a cost analysis of previous disasters could be created for use after large a disaster occurs. A contingency fund could be funded at a level sufficient for large disasters, while relatively routine disasters would still be funded through the DRF. Unlike a rainy-day fund—which pays for disasters through savings and revenue generating measures—a contingency fund would receive an annual appropriation. As the contingency fund is drawn down, the account would be replenished through regular appropriations, obviating the need for emergency supplemental appropriations. During congressional testimony, James

Lee Witt, who was the Director of FEMA during the Clinton Administration, summarized the fund as follows:

> What we did and what worked so well with Congress and with OMB was we were able to use that five year cost analysis to set up a contingency fund that was under OMB that was budgeted, that you did not have to go back for supplementals every time you had a disaster declaration. And then all we would do then is go to OMB and say we have a Presidential disaster declaration that's going to cost $400 million.[46]

- **Model Federal Disaster Funding on State Statutes.** In the majority of states, the first recourse for funding disaster relief is to reallocate departmental appropriations to meet emergency needs. Should the reallocation prove insufficient, the legislature may be called into special session to approve the appropriation of additional funds. In some states, local agency funds must be exhausted before state financial assistance is requested.

In other states, state statutes impose limits on the amount that may be expended on a given disaster. For example, the governor of Alaska may use up to $1 million in state funds per year for disaster assistance. Another example is Hawaii, which caps the amount of funding for a single disaster at $1 million.[47]

Proponents of applying the model used by some states may claim that such measures would reduce the amount needed to fund the DRF. Opponents of the measure, on the other hand, may claim that reallocating funds from other accounts would jeopardize programs that already struggle with tight budgets. They may also argue that should the reallocation of funds for disaster relief become a standard practice, there would be a need to fund potential donor accounts at higher levels, defeating the purpose of reallocating funds to save money. Finally, they may argue that its usage would be impractical for large events such as Hurricane Katrina, which would still require an emergency supplemental appropriation for disaster relief.

If there is interest in changing current practices in funding disaster relief, Members of Congress may wish to further explore other policies for reforming the disaster funding budgeting process such as the ones discussed above, or formulate other strategies that may potentially reform federal disaster relief funding.

## Additional Approaches for Funding Federal Disaster Relief

Should Congress become concerned that disaster relief negatively affects the federal deficit, or that the federal share of disaster relief has become disproportionate, other avenues exist for decreasing the amount of federal funding. Two of these include the following:

- **Decrease the Federal Cost Share.** Generally, when the President declares a disaster the federal government reimburses 75% of a state's disaster relief expenses. The cost share could be altered so that states pay for more of their disaster activities. For example, prior to the Midwest floods in 1993, the federal cost share for mitigation was 50%. Reducing the federal cost share could significantly decrease the amount the federal government pays for disaster relief.
- **Strengthen Declaration Criteria.** It may be argued that the current declaration process allows what some call "marginal" disasters to receive federal funding. Amarginal disaster is an event that some argue could be handled by the state without federal assistance. A set of standardized criteria could be developed to ensure that only events meeting a certain criteria would be eligible for federal disaster relief. Proponents would further argue that scaling back federal fundingmay be beneficial because states may trim investments in mitigation if federal assistance is too generous.[48]

## Oversight on Reporting

A 2006 Government Accountability Office (GAO) report indicated there was a need to improve the information in FEMA's weekly reports on the status of hurricane relief, and that OMB should take action to improve transparency and accountability regarding the status of hurricane related funding.[49] Both OMB and FEMA agreed that these improvements were needed and would be forthcoming. Congress could authorize oversight mechanisms to investigate the extent to which FEMA has made such improvements. For example, Section 203 of H.R. 5351 (introduced in the 109th Congress) would have required each state, local, tribal, and non-profit entity that received federal assistance funds in response to catastrophic events or other emergencies to report to the pertinent federal agency six months after the initial disbursement of resources. Furthermore, the legislation would have also required any agency that disbursed federal assistance funds to report to the Inspector General of the Department the purpose for which resources were provided, the amounts disbursed, allocated, and expended, and the status of reporting by agencies that received disbursements.

## CONCLUDING POLICY QUESTIONS

Since the 1950s, the level of financial assistance given to states for disaster relief by the federal government has steadily increased. In light of the federal deficit, the increased federal involvement has raised policymaking questions concerning how disaster relief should be equitably funded. Some of these questions include the following:

- The model for emergency and disaster response is built on the premise that emergencies and disasters are local. Requests for assistance from the next level of government are made only if that unit of government is overwhelmed. Some would argue that some incidents funded by the federal government do not meet this requirement. An example might be snow removal or repairs after minor flooding. Is the federal government funding emergencies and major disasters that do not meet the criterion of the states being overwhelmed before requesting assistance? Are states using federal funding for disaster relief to protect their budgets?
- Should federal disaster relief be subject to thresholds and maximums? For example, an emergency or major disaster might not receive federal funding unless damage estimates reach a certain level. As another example, the total amount of federal relief for an event could be capped at a certain amount. After this level has been reached, the state would then be responsible to pay for the rest of recovery.
- Should the state's fiscal capability factor into disaster relief? In 1986 FEMA proposed measures to reduce the amount the agency contributed toward disaster relief. One of the proposals argued that funding allocations should be made according to each state's ability to fund its own disaster relief. The determination would be based on a comparison of the state's per capita income with the national per capita income.[50] The calculation would then be used to create a sliding scale for assistance. Communities capable of funding their own disaster relief would receive limited or no assistance. In contrast,

struggling communities would be eligible to receive more federal assistance.[51]
- Some may argue that federal funding for disaster relief has become entrenched to the point that it has contributed to unintended consequences. For example, it has been argued that some states do not properly fund mitigation measures because there is a presumption that federal funding is virtually guaranteed should an emergency or major disaster occur.[52] Those advocating this position could arguably point out that federal involvement in disaster relief will continue to increase and that in order to be fiscally responsible, changes should be made in the way in which disaster relief is funded. Others may claim the function of the federal government is to help states in their time of crisis. Withholding, or limiting the amount of funding a state could receive for an incident would be neglectful of that state's needs.
- OMB's *A New Era of Responsibility* projects that spending for disaster costs for FY2010 will be $11 billion. By 2019 disaster costs are projected to rise to $30 billion.[53] This represents an increase of 173% in disaster costs. On what basis did the Administration calculate this increase in disaster costs?
- Congress requires FEMA to submit a monthly status report on the DRF.[54] The reports must detail obligations, allocations, and expenditures for Hurricanes Katrina, Rita, and Wilma. Other than the DRF report, scant data exist on other federal funding for emergencies and major disasters. In light of the amount of federal funding going to these incidents, could better transparency be achieved by requiring the same level of reporting for all declared emergencies and major disasters?

These and other questions may be raised should Congress elect to debate the past and future funding of disaster relief.

## End Notes

[1] 42 U.S.C. 5121 et seq. For information on the declaration process, see CRS Report RL34146, *FEMA's Disaster Declaration Process: A Primer*, by Francis X. McCarthy.
[2] Per 42 U.S.C. 5122(4), the District of Columbia, the U.S. territories, and tribal governments use the same declaration process.
[3] U.S. Federal Emergency Management Agency. "The Disaster Process and Disaster Aid Programs." http://www.fema.gov/hazard/dproc.shtm.

[4] A structure does not have to be damaged to be eligible for mitigation. Hazard Mitigation Grants are available as block grant funds and can be used outside of the disaster area. For more information, see CRS Report R40471, *FEMA's Hazard Mitigation Grant Program: Overview and Issues*, by Natalie Keegan.
[5] 44 C.F.R. § 206.101(2).
[6] U.S. Congress, Senate Task Force on Funding Disaster Relief, *Bipartisan Task Force on Funding Disaster Relief*, 104th Cong., 1st sess., March 15, 1995, No. 104-4 (Washington: GPO, 1995), p. 1.
[7] 62 Stat. 1031.
[8] See 42 U.S.C. 5202 (note).
[9] P.L. 100-707, 102 Stat. 4708.
[10] Keith Bea, "The Formative Years: 1950-1978," in *Emergency Management: The American Experience*, ed. Claire B. Rubin (Fairfax, VA: Public Entity Risk Institute, 2007), p. 81.
[11] Richard Sylves, *Disaster Policy and Politics: Emergency Management and Homeland Security* (Washington: CQ Press, 2008), p. 48.
[12] P.L. 81-920.
[13] P.L. 81-875, 64 Stat. 1109.
[14] P.L. 89-796, 80 Stat. 1316.
[15] P.L. 93-288, 88 Stat. 143.
[16] 102 Stat. 4689.
[17] Richard Sylves, *Disaster Policy and Politics: Emergency Management and Homeland Security* (Washington: CQ Press, 2008), p. 49.
[18] Richard Sylves, *Disaster Policy and Politics: Emergency Management and Homeland Security* (Washington DC: CQ Press, 2008), p. 54.
[19] Section 1110 the Sandy Recovery Improvement Act of 2013 (P.L. 113-2) amended Sections 401 and 501 of the Stafford Act to allow tribal governments to request disaster assistance from the federal government. Prior to the act the state had to request assistance. For more information on the Sandy Recovery Improvement Act of 2013 see CRS Report R42991, *Analysis of the Sandy Recovery Improvement Act of 2013*, by Jared T. Brown, Francis X. McCarthy, and Edward C. Liu.
[20] CRS Report RL33053, *Federal Stafford Act Disaster Assistance: Presidential Declarations, Eligible Activities, and Funding*, by Francis X. McCarthy.
[21] Title VI of P.L. 109-295, the FY2007 DHS appropriations legislation.
[22] §528, P.L. 109-295, 120 Stat 1383.
[23] A similar reporting requirement has been carried forward in subsequent appropriations statutes. See, for example, P.L. 110-329, 122 Stat 3674-75.
[24] This section is based on the author's in-person interview with a Department of Homeland Security (DHS) finance official, March 9, 2009.
[25] The DRF is a "no year" account; funds remain until expended.
[26] An example of a duplication of a benefit fund is when the state receives an insurance payment for disaster damages after obtaining federal funding for the same damages. In such a case the state is to return the funding to the federal government.
[27] Open incidents are those whose related activities are currently being funded by FEMA.
[28] Based on U.S. Government Accountability Office, *Disaster Relief: Government Framework Needed to Collect and Consolidate Information to Report on Billions in Federal Funding for the 2005 Gulf Coast Hurricanes*, GAO-06-834, September 2006. The Department of Defense (DOD) is the largest recipient of supplemental funding.
[29] Office of Management and Budget, *A New Era of Responsibility: Renewing America's Promise*, Washington DC, February 26, 2009, p. 36.
[30] For more information on rescissions and disaster relief see CRS Report R42458, *Offsets, Supplemental Appropriations, and the Disaster Relief Fund: FY1990-FY2013*, by William L. Painter.

[31] U.S. Congress, Congressional Budget Office, *Supplemental Appropriations in the 1990s*, March 2001, http://www.cbo.gov/ftpdocs/27xx/doc2768/EntireReport.pdf.

[32] CQ Financial Transcripts, "OMB Director Orszag and CEA Chair Romer Hold Briefing on the Fiscal 2010 Budget Request," press release, February 26, 2009, http://www.cq.com/display.do?dockey=/cqonline/prod/data/docs/html/transcripts/financial/111/financialtranscripts111-000003061915.html@committees&pub=financialtranscripts&print=true#speakers. The President's request for the DRF for FY2010 was increased by $600 million over the FY2009 enacted amount of $1.4 billion.

[33] For more information on supplemental funding for Hurricane Sandy see CRS Report R42869, *FY2013 Supplemental Funding for Disaster Relief*, coordinated by William L. Painter and Jared T. Brown. See also CRS Report R42803, *Federal Involvement in Flood Response and Flood Infrastructure Repair: Hurricane Sandy Recovery*, by Nicole T. Carter.

[34] Congress did not meet the full request of the American Red Cross, which requested $150 million for reimbursement of disaster relief and recovery expenditures as a result of disasters occurring in 2008. This is not the first time Congress appropriated funds for the organization. In 2004, Congress gave $70 million in aid to the American Red Cross after four hurricanes hit Florida (118 Stat, 1251-1252).

[35] For information on federal expenditures for Hurricanes Katrina, Rita, Wilma, Gustav, and Ike see CRS Report R43139, *Federal Disaster Assistance after Hurricanes Katrina, Rita, Wilma, Gustav, and Ike*, coordinated by Bruce R. Lindsay and Jared C. Nagel.

[36] See CRS Report RL33188, *Protecting New Orleans: From Hurricane Barriers to Floodwalls*, by Nicole T. Carter.

[37] Some of the funding in P.L. 110-329 was directed at the 2005 hurricanes and hurricanes, flooding, and other disasters occurring in 2008. An exact amount for each of these events could not be identified from the legislative text.

[38] See CRS Report RL34410, *The Louisiana Road Home Program: Federal Aid for State Disaster Housing Assistance Programs*, by Natalie Keegan.

[39] In requests to Congress, President Bush termed the sequence of events as a "reallocation" of funds.

[40] See **Table 3**.

[41] For more information on disaster assistance expenditures for Hurricanes Katrina, Rita, Wilma, as well as information on Gustav and Ike CRS Report R43139, *Federal Disaster Assistance after Hurricanes Katrina, Rita, Wilma, Gustav, and Ike*, coordinated by Bruce R. Lindsay and Jared C. Nagel.

[42] CRS Report RS20371, *Overview of the Authorization-Appropriations Process*, by Bill Heniff Jr.

[43] A ceiling arguably limits the amount of funding an account may receive. The ceiling however, is not binding, Congress may appropriate funds at levels above the authorization.

[44] 42 U.S.C. 5197e.

[45] U.S. Congress, Senate Bipartisan Task Force on Funding Disaster Relief, *Federal Disaster Assistance*, Report of the Senate Task Force on Funding Disaster Relief, 104th Cong., 1st sess., March 15, 1995, No. 104-4 (Washington: GPO, 1995), pp. 69-74.

[46] U.S. Congress, House Committee on Small Business, *Hearing on Disaster Relief and Access to Capital Legislation*, 110th Cong., 1st sess., March 8, 2007, 110-6.

[47] CRS Report RL32287, *Emergency Management and Homeland Security Statutory Authorities in the States, District of Columbia, and Insular Areas: A Summary*, by Keith Bea, L. Cheryl Runyon, and Kae M. Warnock.

[48] For more information on changing declaration criteria see CRS Report R42702, *Stafford Act Declarations 1953-2011: Trends and Analyses, and Implications for Congress*, by Bruce R. Lindsay and Francis X. McCarthy. See also James F. Miskel, *Disaster Response and Homeland Security: What Works, What Doesn't* (Westport, CT: Praeger Security International, 2006), p. 126.

[49] U.S. Government Accountability Office, *Disaster Relief: Governmentwide Framework Needed to Collect and Consolidate Information to Report on Billions in Federal Funding for the 2005 Gulf Coast Hurricanes*, GAO-06-834, September 2006.

[50] U.S. Congress, House Committee on Public Works and Transportation, Subcommittee on Investigations and Oversight, *The Federal Emergency Management Agency's Proposed Disaster Relief Regulations (Budget Driven Rulemaking)*, committee print, 100$^{th}$ Cong., 1$^{st}$ sess., August 1987, 75-963 (Washington: GPO, 1987), pp. 4-5.

[51] Under current law (42 U.S.C. 5163) areas cannot be precluded from receiving assistance *solely* on the basis of a sliding scale or formula. Congress amended the Stafford Act in this fashion as a response to the 1986 proposed regulation.

[52] See James F. Miskel, *Disaster Response and Homeland Security: What Works, What Doesn't* (Westport, CT: Praeger Security International, 2006), or Rutherford H. Platt, *Disasters and Democracy: The Politics of Extreme Natural Events* (Washington DC: Island Press, 1999) for arguments against increased federal involvement in disaster assistance.

[53] Office of Management and Budget, *A New Era of Responsibility: Renewing America's Promise*, Washington DC, February 26, 2009, p. 117.

[54] P.L. 110-161, also 42 U.S.C. 5208.

In: Federal Disaster Assistance ...
Editor: Madeline Payton

ISBN: 978-1-63117-886-3
© 2014 Nova Science Publishers, Inc.

*Chapter 3*

# FEMA'S DISASTER DECLARATION PROCESS: A PRIMER[*]

## *Francis X. McCarthy*

### SUMMARY

The Robert T. Stafford Disaster Relief and Emergency Assistance Act (referred to as the Stafford Act - 42 U.S.C. 5721 et seq.) authorizes the President to issue "major disaster" or "emergency" declarations before or after catastrophes occur. Emergency declarations trigger aid that protects property, public health, and safety and lessens or averts the threat of an incident becoming a catastrophic event. A major disaster declaration, issued after catastrophes occur, constitutes broader authority for federal agencies to provide supplemental assistance to help state and local governments, families and individuals, and certain nonprofit organizations recover from the incident.

The end result of a presidential disaster declaration is well known, if not entirely understood. Various forms of assistance are provided, including aid to families and individuals for uninsured needs and assistance to state and local governments and certain non-profits in rebuilding or replacing damaged infrastructure.

The amount of assistance provided through presidential disaster declarations has exceeded $140 billion. Often, in recent years, Congress has enacted supplemental appropriations legislation to cover

---

[*] This is an edited, reformatted and augmented version of a Congressional Research Service publication, CRS Report for Congress RL34146, from www.crs.gov, dated May 18, 2011.

unanticipated costs. While the amounts spent by the federal government on different programs may be reported, and the progress of the recovery can be observed, much less is known about the process that initiates all of this activity. Yet, it is a process that has resulted in an average of more than one disaster declaration a week over the last decade.

The disaster declaration procedure is foremost a process that preserves the discretion of the governor to request assistance and the President to decide to grant, or not to grant, supplemental help. The process employs some measurable criteria in two broad areas: Individual Assistance that aids families and individuals and Public Assistance that is mainly for repairs to infrastructure. The criteria, however, also considers many other factors, in each category of assistance, that help decision makers assess the impact of an event on communities and states.

Under current law, the decision to issue a declaration rests solely with the President. Congress has no formal role, but has taken actions to adjust the terms of the process. For example, P.L. 109-295 established an advocate to help small states with the declaration process. More recently, Congress introduced legislation, H.R. 3377, that would direct FEMA to update some of its criteria for considering Individual Assistance declarations.

Congress continues to examine the process and has received some recommendations for improvements. Given the importance of the decision, and the size of the overall spending involved, hearings have been held to review the declaration process so as to ensure fairness and equity in the process and its results.

## BACKGROUND

Under the Robert T. Stafford Disaster Relief and Emergency Assistance Act (P.L. 93-288) there are two principal forms of presidential action to authorize federal supplemental assistance. Emergency declarations are made to protect property and public health and safety and to lessen or avert the threat of a major disaster or catastrophe.[1] Emergency declarations are often made when a threat is recognized (such as the emergency declarations for Hurricane Katrina which were made prior to landfall) and are intended to supplement and coordinate local and state efforts prior to the event such as evacuations and protection of public assets. In contrast, a major disaster declaration is made as a result of the disaster or catastrophic event and constitutes a broader authority that helps states and local communities, as well as families and individuals, recover from the damage caused by the event.[2] The differences between the two forms of declarations remain an area of study regarding what events may or may not qualify for the respective declarations.[3]

Federal disaster assistance has served as the impetus for many supplemental appropriations bills over the last several decades, and has accounted for presidential declarations in every state and territory.[4] The supplemental funds are placed in the Disaster Relief Fund (DRF) which is a "no-year" fund managed by the Federal Emergency Management Agency (FEMA) and used only for spending related to presidentially declared disasters.

Major disasters can be a dominant story in the mass media that captures attention both for the devastation that results as well as the potential help that is expected. As one observer noted:

> Disaster assistance is an almost perfect political currency. It serves humanitarian purposes that only the cynical academic could question. It is largely funded out of supplemental appropriations and thus does not officially add to the budget deficit. It promotes the local economy of the area where the building process occurs.[5]

While disaster assistance may be good "political currency," a disaster declaration is generally the result of a tragic and devastating incident that disrupts (and sometimes takes) the lives of hundreds or thousands of families and individuals and the communities and states where they reside. The long-term economic and environmental impact of a disaster can be severe. The assistance offered from federal and private sources may or may not be commensurate with the damage inflicted by a natural or man-made event. Following a disaster, years of rebuilding and recovery work may lie ahead for communities and states. It is the declaration process that sets the federal recovery help in motion.

The trigger for federal disaster assistance is contained in a relatively short statutory provision. P.L. 93-288 (the Stafford Act) includes one brief section that establishes the legal requirements for a major disaster declaration:

> Section 401. Procedures for Declaration. All requests for a declaration by the President that a major disaster exists shall be made by the Governor of the affected state. Such a request shall be based on a finding that the disaster is of such severity and magnitude that effective response is beyond the capabilities of the state and the affected local governments and that the federal assistance is necessary. As a part of such request, and as a prerequisite to major

disaster assistance under this Act, the Governor shall take appropriate response action under state law and direct execution of the state's emergency plan. The Governor shall furnish information on the nature and amount of State and local resources which have been or will be committed to alleviating the results of the disaster and shall certify that, for the current disaster, state and local government obligations and expenditures (of which state commitments must be a significant proportion) will comply with all applicable cost-sharing requirements of this Act. Based on the request of a Governor under this section, the President may declare under this Act that a major disaster or emergency exists.[6]

The process for an emergency declaration is also contained in the Stafford Act. In part it is similar to a major disaster declaration in finding and process, but the actual authorities are limited. In addition, section (b) provides for a special authority for the President to exercise his discretion for events that have a distinctly federal character:

Sec. 5191. Procedure for declaration. (a) Request and declaration All requests for a declaration by the President that an emergency exists shall be made by the Governor of the affected State. Such a request shall be based on a finding that the situation is of such severity and magnitude that effective response is beyond the capabilities of the State and the affected local governments and that Federal assistance is necessary. As a part of such request, and as a prerequisite to emergency assistance under this chapter, the Governor shall take appropriate action under State law and direct execution of the State's emergency plan. The Governor shall furnish information describing the State and local efforts and resources which have been or will be used to alleviate the emergency, and will define the type and extent of Federal aid required. Based upon such Governor's request, the President may declare that an emergency exists. (b) Certain emergencies involving Federal primary responsibility. The President may exercise any authority vested in him by section 5192 of this title or section 5193 of this title with respect to an emergency when he determines that an emergency exists for which the primary responsibility for response rests with the United States because the emergency involves a subject area for which, under the Constitution or laws of the United States, the United States and authority. In determining whether or not such an emergency exists, the President shall consult the Governor of any affected State, if practicable. The President's determination may be made without regard to subsection (a) of this section.[7]

The declaration process is elaborated upon in regulations, specifically in Subpart B of Part 206 (44 CFR). While these regulations have been adjusted through the regulatory process during the past three decades, since 1974 the procedures have undergone little significant change. The process itself is representative of the historical progression of federal disaster relief from being of an episodic nature to the current commonplace disaster declaration, now occurring on a weekly basis. The context in which disaster relief has grown has been in keeping with the growth of government and its concerns.

> Federal disaster relief has a long history in the U.S. dating back to the last years of the eighteenth century and arguably provided much of the political genesis for the New Deal social welfare programs (Landis 1999; Landis 1998; Moss 1999). As Michele Landis argues, social and political construction of claimants for relief as helpless victims of external forces beyond their control ("Acts of God") have exerted an enduring influence on American political discourse, which has manifested itself in heavy reliance on prior political precedents and analogies in constructing responses to current disasters.[8]

Information contained in the Preliminary Damage Assessments (PDAs) and summaries prepared by FEMA regional offices that accompany gubernatorial requests were long considered "pre-decisional and deliberative information" by the executive branch because they are part of the package that is developed and sent to the White House for the President's review and ultimate decision. These materials have generally not been available under the Freedom of Information Act process.[9] However, during 2008, FEMA began to list on its website the results of the PDAs. While the summaries of FEMA recommendations are still not available, the agreed upon figures from the PDAs are now available to the public for review.[10]

Policy makers have found it difficult to achieve equity in the treatment of disparate natural disaster events. The events can vary widely in their type, scope, duration, and impact. Perhaps the greatest variables are the states they affect. Each state has a different topography, a different history, and different capacities to respond and recover based on their own authorities, resources, and choices in what they will do following a disaster. Given those variations, it is a daunting task to construct a uniform process that can account for the range of natural and governmental circumstances that are a part of the nation's potential disaster landscape.

## CONGRESS AND THE DECLARATION PROCESS

### The Impetus for Reform

The question of how the federal disaster declaration process should work has come to the attention of Congress from time to time. Congress has requested reports from its investigative arms and investigated the process in panels that have considered aspects of the disaster relief process.[11] Various administrations have also considered the process and FEMA's role in it.

In 1981, and again in 2001, GAO issued reports on the declaration process that questioned the quality and consistency of FEMA's assessment criteria as well as the agency's ability to produce valid recommendations to the President on a governor's request for supplemental aid. As the most recent report (2001) concluded:

> These criteria are not necessarily indicative of a state's ability to pay for the damage because they do not consider the substantial differences in states' financial capacities to respond when disasters occur. As a result, federal funds may be provided for some disasters when they are not needed – a result that would be inconsistent with the Stafford Act's intent.[12]

Congressional interest in the declaration process derived, in part, from the increased cost in emergency spending—a recurring subject for the appropriations committees when considering supplemental spending legislation.[13] One review of disaster funding, particularly supplemental appropriations, found that the great majority of Disaster Relief Fund spending is related to the large, catastrophic events that meet the criteria established in the Stafford Act. These supplementals have been "driven" by the urgency of large natural disasters. As one witness explained in congressional testimony:

> Since funds provided in annual appropriations measures may be used to cover the federal costs of less catastrophic major disasters, Congress has focused critical attention on "must pass" legislation that would replenish the Disaster Relief Fund for the worst and most costly disasters.[14]

A review of data for a seven-year period from 1988 to 1995 reveals that large expenditures, as funded by supplemental bills, relate to declarations issued for the largest events.[15] During this time period, disaster declarations were made for Hurricane Hugo, the Loma Prieta earthquake, Hurricane Andrew, the Midwest floods of 1993, and the Northridge earthquake. However, these were not the only events deemed worthy of presidential action and of cost to the federal treasury. As summarized by one author:

> But like the tail of a comet, over 200 other declarations accounted for one quarter of such outlays, many of them of relatively minute cost and extent. While of lesser impact on the national treasury, such "low end" declarations have become, to some observers, new sources of federal spending at the local level, long referred to in other contexts as "pork barrel spending."[16]

During early 2001, the term "entitlement" was beginning to be used in describing federal disaster spending by the new Bush Administration. Underlining this thinking in his first appearance before the Senate Appropriations Committee, FEMA Director Joe M. Allbaugh explained:

> FEMA is looking at ways to develop a meaningful and objective criteria for disaster declarations that can be applied consistently. These criteria will not preclude the President's discretion but will help states better understand when they can reasonably turn to the federal government for assistance and when it would be more appropriate for the state to handle the disaster itself.[17]

During the 110th Congress, attention was given to the disaster declaration process and what states and local governments can reasonably expect from that process. During the early spring of 2007, there were tornadoes that had an impact on Arkansas, Alabama and Georgia. The latter two states received disaster declarations while the damage in Arkansas was deemed insufficient to warrant federal assistance. Following these actions, Chairman Bennie Thompson of the House Homeland Security Committee scheduled a hearing to review FEMA's procedures. Representative Thompson set the context as follows:

> As Members representing real communities back home, we want to understand just how FEMA makes determinations regarding what is a disaster deserving of attention and when folks have to fend for themselves. Quite simply, we must have a serious discussion on what our expectations are of our federal government and what should remain a state and local responsibility.[18]

In order to examine the process and the financial projections of disaster spending in particular, Congress mandated in P.L. 110-28 that GAO study how FEMA develops its estimates for supplemental funding that may be needed.[19] Report language detailed these instructions for the review:

> The Committee continues to be concerned with FEMA's ability to manage resources in a manner that maximizes its ability to effectively and efficiently deal with disasters. One aspect of particular concern is how FEMA makes projections of funding needed in response to any given disaster or to meet future disasters. A recent Government Accountability Office (GAO) report raised concerns about FEMA's ability to manage its day-to-day resources and the lack of information on how FEMA's resources are aligned with its operations. As a follow-up to this report, the Committee requests that within six months of enactment GAO review how FEMA develops its estimates of the funds needed to respond to any given disaster. Such review should include how FEMA makes initial estimates, how FEMA refines those estimates within the first few months of a disaster, and how closely FEMA's estimates predict actual costs. The review should also include additional analysis and recommendations regarding FEMA's ability to manage disaster-related resources in a manner that maximizes effective execution of its mission.[20]

The study parameters mandated by Congress are broader than the declaration process, placing its emphasis on ongoing FEMA spending, including the refinement and revision of early estimates as well as the general management of spending projections for the Disaster Relief Fund (DRF). But it recognizes the importance of the initial estimates in decision-making and that the declaration process represents the fundamental decision to both establish federal participation following a disaster and provides the initial estimated amount of resources needed that informs that decision.

GAO's eventual report in response to this request suggested that FEMA's estimates for disaster costs have improved but could use additional refinement; for example

> some of the factors that can lead to changes in FEMA's cost estimates are beyond its control, such as the discovery of hidden damage. Others are not, such as its management of mission assignments. Sensitivity analyses to identify the marginal effect of key cost drivers could provide FEMA a way to isolate and mitigate the effect of these factors on its early estimates. To better predict

applicant costs for the Individual Assistance program, FEMA could substitute or add more geographically specific indicators for its national average.[21]

During the 111th Congress, a hearing before the House Homeland Security Committee focusing on FEMA's regional offices also elicited comments regarding the relative speed, and perceived lack of transparency, of the declaration process. The comments came from both a committee member and a state official closely involved in the process.

> Ranking Republican Mike D. Rogers of Alabama questioned the timeliness of FEMA's disaster declarations. Declarations that should take a few days often take a month or longer, and a lack of available information on the status of requests only adds to state officials' frustration, said Brock Long, director of Alabama's Emergency Management Agency. "When we make phone calls to the region or to headquarters, a lot of time the answer we get is 'the declaration request is in process' and that's it," Long said.[22]

## The Skepticism of Reform

There have been several proposals to impose more stringent regulations on the declaration process in an attempt to more precisely assess disaster impacts, calculate eligible damage, and incorporate some measure of suffering and loss. Differing perceptions of the declaration process resulted in different reactions to these proposed reforms.

The history of the disaster declaration process is rife with reform efforts that were perceived by some not as reform but as punitive measures directed at certain constituencies. Some of those differing perceptions were held by those closest to the event at the local level that had experienced a disaster. A different perception was also often held by governors that wanted to protect their option to request federal help. These perceptions were also shared, in some instances, by Members of Congress representing affected areas. All elected officials argued, in various forms, that reform should not impede the delivery of needed federal aid.

One example, now in law, of the desire to reform but not obstruct the declaration process is in the Post-Katrina Emergency Management Reform Act of October of 2006. That statute created a new position at FEMA: the Small State and Rural Advocate.[23] One of the principal duties of the advocate is to ensure that the needs of smaller states and rural communities will be "met in the declaration process;" the advocate is also directed to "help small states prepare declaration requests, among other duties."[24]

Another example was FEMA's response to recommendations from GAO's 1981 report. FEMA drafted regulations in 1986 that would be more certain in their delineation of state requirements prior to a declaration, reduce overall federal contributions, and install a formula to determine whether a state would receive a presidential disaster declaration for repairs to state and local infrastructure (known as "public assistance" in the Stafford Act). As the congressional report on the initiative summarized the FEMA draft:

> The proposal would have limited the number of future presidential declarations by establishing a "state deductible" based on a per capita minimum dollar amount adjusted by the ratio of the state/local price index to the national index. Of 111 declarations issued in prior years, 61 would have been ineligible for any public assistance. FEMA also proposed to decrease the federal share for disaster costs from 75 percent to 50 percent and to exclude aid to special districts.[25]

The FEMA proposal of 20 years ago addressed some problems identified by students and critics of the declaration process. However, some Members of Congress viewed this proposal as a means of removing the power of discretion from elected leadership. Rather than apply the empirical solution suggested by FEMA to perceived problems in the declaration process, Congress instead legislated a provision to explicitly forbid the primacy of any "arithmetic formula." In place of agreeing to the regulatory changes in the formula proposed by FEMA and the Administration, Congress added the following section to the Stafford Act:

> Limitation on the Use of Sliding Scales. Section 320. No geographic area shall be precluded from receiving assistance under this Act solely by virtue of an arithmetic formula or sliding scale based on income or population.[26]

While enactment of this legislation halted FEMA's efforts, the issue did not disappear. As discussed below, FEMA has since adopted regulations that use, but not "solely", arithmetic formulae in determining need for assistance.

## PRESIDENTIAL AND GUBERNATORIAL DISCRETION

The declaration process contains many factors for consideration and, for all but the most catastrophic events, the process moves at a deliberate speed accumulating information from several sources. While the process is informed by that information and its relationship to potential assistance programs, the information that is gathered at the state and local level does not preclude the exercise of judgment by the governor or the President.

The Stafford Act stipulates several procedural actions a governor must take prior to requesting federal disaster assistance (including the execution within the state of the state emergency plan and an agreement to accept cost-share provisions and related information-sharing). Still, the process leaves broad discretion with the governor if he or she determines that a situation is "beyond the capabilities of the state." The concession that a state can no longer respond on its own is difficult to quantify. It is the governor who makes that assessment, based on his or her knowledge of state resources and capabilities.

Actions by a governor are a driving constant in this process. Both declarations of major disaster and declarations of emergency must be triggered by a request to the President from the governor of the affected state.[27] The President cannot issue either an emergency or a major disaster declaration without a gubernatorial request. The only exception to this rule is the authority given to the President to declare an emergency when "he determines that an emergency exists for which the primary responsibility for response rests with the United States because the emergency involves a subject area for which, under the Constitution or laws of the United States, the United States can exercise exclusive or preeminent responsibility and authority."[28]

The importance of governors in the process is not lost on, nor would it likely be diminished by, Presidents who formerly served in that position. Over the last three decades of Stafford Act implementation, four of the Presidents during this period were former governors who had worked through the disaster declaration process from both the state and the federal level.[29]

Having that experience may have left the Presidents, and their staffs and appointees, with an appreciation of the discretionary authority inherent in the process. While there are some established standards in the law, the factors that are to be weighed in considering the impact of a disaster on the need for assistance for families and individuals are general considerations that underscore the judgment required to reach a decision. As one observer noted of the decision for general, flexible considerations:

In other words, Congress likes to keep the process imprecise, even if benefits occasionally go to the undeserving. The absence of objective criteria preserves wide political discretion to the president.[30]

Congress also has among its number former governors who have exercised this discretion at the state level. Still, despite the interest some may have in keeping "the process imprecise," some Members of Congress express disappointment at times with the exercise of the discretion and the general nature of the considerations. As noted previously in this report, that disappointment has been reflected in hearings that have focused on how the disaster declaration process works in practice.

In the declaration process, FEMA develops a recommendation that is sent to the White House for action. However, as implied, it is a recommendation from FEMA and the Department of Homeland Security (DHS). The final action, as defined in the regulations of the process, is a "Presidential determination."[31] Just as the governor retains the discretion to request federal assistance regardless of thresholds or indicators, the President retains the discretion to make a decision that may be counter to recommendations he receives.

## PRELIMINARY DAMAGE ASSESSMENTS

Although not explicitly mentioned in the Stafford Act, Preliminary Damage Assessments (PDAs) are a crucial part of the process of determining if an event may be declared a major disaster by the President. The minimal discussion of PDAs in the public record stands in inverse proportion to their impact on disaster decisions and subsequent expenditures from the Disaster Relief Fund (DRF).

When a PDA is conducted after an event it is the "mechanism used to determine the impact and magnitude of damage and the resulting unmet needs of individuals, businesses, the public sector, and the community as a whole."[32] The most "preliminary" part of a PDA may be an abbreviated one completed only by the state to determine if the situation merits development of a complete PDA with federal participation. Based on their previous experience, states may determine that the event will not reach the level where a federal disaster declaration is likely. However, despite findings that federal aid may not be needed, there may be political considerations that could lead to a gubernatorial request. As one author points out in his study of the process:

> Governors also feel the heat of media coverage of incidents in their states, and they too appreciate the importance of exhibiting political responsiveness. Governors also appreciate that their future political fortunes may be influenced by how they handle their disaster and emergency incidents. As a consequence, the hypothesis assumes that governors are the pivotal and decisive players in securing presidential disaster declarations and that they have a tendency to request declarations for even marginal events.[33]

While media and political pressure may have some influence on the outcome of some requests, governors may exercise caution since they are reluctant to be turned down when requesting aid. A denial of their request could be perceived by some to reflect adversely on their decision-making skills and judgment under pressure. Unlike the procedures of the federal process, a governor's decision to request a declaration is a public and often newsworthy action.

Regardless of any question regarding motivation, the governor's first decision is whether the incident is severe enough to assemble a traditional PDA team to survey the damaged area. The traditional PDA team includes a state official, representatives from the appropriate FEMA regional office, a local official familiar with the area and, in some instances, representatives from the American Red Cross and/or the Small Business Administration. The FEMA representatives have the responsibility of briefing the team on the factors to be considered, the information that will be helpful in the assessment and how the information should be reported. One significant improvement in this process is that the regulations now require that the participants reconcile any differences in their findings.[34]

Another factor is the quality of the PDA team and its findings. PDAs are ordered up quickly after an event. FEMA's 10 regional offices are often engaged in multiple disasters and have to rely on temporary employees (albeit, usually experienced ones) to staff the PDA team.[35] Also, given the variables among regions in interpreting policy and guidance, consistency of approach may also be a question. This became a significant issue during the hurricane season of 2004 in Florida, where questions were raised regarding the designations of some counties.[36]

When working openly and with federal and state cooperation, the PDA can be an effective and inclusive process. As Washington State Emergency

Management Director Jim Mullen, and FEMA Region X Director Susan Reinertson, explained the process for a survey of flood damage in November of 2006:

> 'Joint PDA teams will visit and inspect damaged areas, document damage and talk, as needed, with homeowners and local officials,' said Mullen. 'It's a partnership effort designed to provide a clear picture of the extent and locations of damage in counties that have reported the most substantial damage to primary homes and businesses.'
> 'The PDA teams look at the total scope of damage to establish if recovery is beyond the capabilities and resources of the state and local governments. The PDA doesn't determine the total cost of recovery, nor does it guarantee a presidential declaration for individual assistance,' said FEMA Regional Director Susan Reinertson.[37]

PDA teams often face challenges in the collection of data. Some information may be observable in a survey of the area, such as the number of bridges damaged or the number of culverts washed out. But other necessary information, such as the percentage of elderly residents in an area, or the amount of insurance coverage for all homeowners or renters may be more difficult to obtain. Also the geographic span of the damage can create complications as the PDA team struggles to cover all of the affected area in a limited time. Further complications may then ensue based on how much of the area that the PDA teams have visited (generally counties) are included in the governor's request.

While PDAs are the usual way damages are assessed, there are exceptions to this rule. Some incidents are so massive in their scale and impact that the actual declaration is not in doubt. This would include events such as Hurricanes Andrew and Katrina, the Loma Prieta and Northridge earthquakes, the Mount St. Helens volcanic eruption, and the September 11 terrorist attacks. In these instances, the decision is not whether a declaration will be made, but how broad the coverage will be, both geographic and programmatic. In such cases the President, in accordance with the regulations, can waive the PDA requirement. But as the regulations note, a PDA may still be needed "to determine unmet needs for managerial response purposes."[38] This means the PDA helps to identify a specific, potential need for certain programs, such as crisis counseling or disaster unemployment assistance during the disaster recovery period. It is this identification of discrete need that helps the governor decide on which assistance programs will be requested.

The PDA is a "bottom up" process as information gradually rises up for decision-makers to consider. Multiple pressure points, including affected citizens, elected officials, and professionals in various fields, may all urge PDA team members to reach certain conclusions. As one author suggests:

> The president, motivated by the need to appear highly politically responsive, solicits and encourages a gubernatorial request for a presidential disaster declaration. Publicity is a factor in that CNN and other news organizations help to promote nationally what would otherwise be a local incident addressed by subnational authorities. The president may also be influenced by the electoral importance of the state that experiences the incident.[39]

It is difficult to overstate the importance of the media context as noted above. Depending on the news of the day, the disaster event in question may be the biggest national story and thus create momentum for action that is difficult to assuage with explanations of traditional administrative procedures.

## FACTORS CONSIDERED FOR PUBLIC ASSISTANCE IN MAJOR DISASTER DECLARATIONS

Public Assistance (PA) refers to various categories of assistance to state and local governments and non-profit organizations. Principally, PA covers the repairs or replacement of infrastructure (roads, bridges, public buildings, etc.) but also includes debris removal and emergency protective measures which cover additional costs for local public safety groups incurred by their actions in responding to the disaster.[40] In assessing the degree of PA damage, FEMA considers six general areas:

- Estimated cost of the assistance
- Localized impacts
- Insurance coverage
- Hazard mitigation
- Recent multiple disasters
- Programs of other federal assistance.

## Estimated Cost of the Assistance

Although all of these factors are considered, the *estimated cost of the assistance* is a key component and may be "more equal" than other factors since it contains a threshold figure. What is especially noteworthy about the cost estimates is that they could be interpreted as an example of the "arithmetic formula" that was precluded from use in Section 320 of the Stafford Act. However, that section does state that a formula cannot be "solely" determinative of the fate of a governor's request. Use of the other factors listed arguably justifies FEMA's consideration of the threshold figure of $1 million in PA damage—the first number FEMA expects to see in a request that includes PA as an area of needed assistance. In addition FEMA also considers a state-wide threshold of $1.30 per capita in estimated eligible disaster costs before it will approve a request for PA help.[41]

Depending on the state's population, the per capita threshold may be difficult to reach. For example, the 2000 Census estimated California's population at just under 34 million people.[42] Applying the $1.30 per capita figure, it would require eligible PA damage in California to be close to $41.5 million. California is a large state with a budget and tax base commensurate with its size. Expecting a large state to be able to respond on its own is equating such help, and such amounts, to be within "the capabilities of the state."[43]

Compare that level of eligible damage for California ($41.5 million) with Nevada, a small population state according to FEMA regulations.[44] For Nevada, with a population of just under 2 million people according to the 2000 Census, eligible PA damage of about $2.4 million would make the state potentially eligible for supplemental federal assistance.

There are obvious differences in the populations of these two states; however, both have substantial industries and are growing areas. They are bordering states that are both subject to the threat of earthquake damage.[45] Different measurements have been suggested to try to more accurately capture a state's capacity to respond to disaster events. Some of these measurements are discussed in the "Congressional Considerations for the Declaration Process" section of this report.

## Localized Impacts

The next factor used by FEMA to consider a declaration that may include PA help focuses on *localized impacts*. FEMA generally looks for a minimum of $3.27 per capita in infrastructure damage in a county before designating it for PA funding.[46] While such a per capita amount would likely ensure that the local entity (almost always a county) would be included in the list of jurisdictions designated for PA assistance should a declaration be issued, high local levels of damage per capita would not supersede a finding that damages state-wide fell below the statewide threshold amount. However, knowledge of a very large localized impact could weigh on the President's discretion and raise the importance of the *localized impact* factor.

## Insurance Coverage

Insurance coverage also is considered in assessing damage. Officials preparing PA estimates deduct the amount of insurance that should have been held by units of governments and nonprofit organizations from the total eligible damage amount. However, this is a complicated assessment with several caveats. A considerable number of states and local governments "self-insure" their investments against some types of disasters and may argue that they would have total liability absent a federal disaster declaration.[47] Also, in the event of some disasters (earthquakes being a prime example), the state insurance commissioner must certify that hazard insurance is both available and affordable. If it is not available, deducting an amount of coverage not realistically available would not be a practical consideration.[48]

In the case of a flood event, it is much simpler for FEMA to access information available on whether flood insurance was available, given that FEMA administers the National Flood Insurance Program (NFIP) and there are program staff in each FEMA region who monitor NFIP participation. This knowledge of flood insurance availability and costs makes it a simpler process for FEMA to deduct the amount of flood insurance that should have been in place from the total potential award for a public structure.[49]

## Hazard Mitigation

Hazard mitigation presents a challenging and different type of consideration for Public Assistance disaster declaration requests. If the requesting state can prove that their per capita amount of infrastructure damage falls short due to mitigation measures that lessened the disaster's impact, FEMA will consider that favorably in its recommendation to the President. Calculations of savings would be dependant on cost-benefit analysis and other related estimates of damages that were avoided. This factor is intended to encourage mitigation projects by states to lessen the risks of future natural disasters. Some might consider this practice to be one where "a good deed should go unpunished." However, another factor to be considered is that the mitigation work that is credited to the state may, in fact, have been principally financed (up to 75% of the costs) with previous FEMA disaster assistance funding through the Hazard Mitigation Grant Program (HMGP).[50]

## Recent Multiple Disasters

Recent multiple disasters is the factor FEMA considers when a state has been repeatedly hit by disaster events (either presidential declarations or events within the state that were not declared) within the previous 12 months. FEMA evaluates the amount of funds that the state has committed to these recent events and their impact on the state and its residents. For example, a request from a state that has responded on it own to a series of tornadoes may receive a more favorable consideration, even if the catalyst for the request was arguably not as destructive as others. This factor was used recently in FEMA's assessment of the multiple hurricanes that struck Florida during the 2004 hurricane season.

## Other Federal Programs

When FEMA is reviewing a governor's request it is also considering whether *other federal programs* are available. One example might be if a large amount of the reported damage occurred to federal-aid-system roads. The Federal Highway Administration (FHWA) is a more appropriate agency to handle such an event, and administers programs that address this specific type of damage.[51] The bridge collapse in Minnesota in August of 2007 was a dramatic example of this type of event that would warrant significant, non-FEMA, federal aid. Similarly, other federal programs may be more responsive to certain types of natural events and the problems they create. For example, oceanic bacteria could cause harmful failures for the fishing and hatcheries industry. Should such outbreaks occur, the Magnuson-Stevens Fisheries Act (P.L. 94-265) authorizes programs under the Commerce Department that would have more appropriate forms of emergency assistance for commercial fishermen than FEMA through the Stafford Act.[52] Also, while drought is a type of disaster that could result in a major disaster, such catastrophes usually result in program assistance from the U.S. Department of Agriculture rather than presidential declarations.[53]

In summary, all of these factors are used by FEMA to determine whether a major disaster declaration will be recommended and whether PA aid will be extended as a part of that potential declaration. FEMA's assistance to a homeowner who chooses not to purchase insurance is left with help from FEMA that is capped at $28,200. The assistance to repair public infrastructure is based on the amount of eligible damage caused by the disaster event. There is no cap, and the sums can be in the billions of dollars.

## FACTORS CONSIDERED FOR INDIVIDUAL ASSISTANCE IN MAJOR DISASTER DECLARATIONS

Individual Assistance (IA) includes various forms of help for families and individuals following a disaster event. The assistance authorized by the Stafford Act can include housing assistance, disaster unemployment assistance, crisis counseling and other programs intended to address the needs of people. In seeking to assess the impact of a disaster on families and individuals, the factors FEMA considers include:

- Concentration of damages
- Trauma
- Special populations
- Voluntary agency assistance
- Insurance coverage
- Average amount of Individual Assistance by state.

## Concentration of Damages

Concentration of damages looks at the density of the damage in individual communities. FEMA's regulations state that highly concentrated damages "generally indicate a greater need for federal assistance than widespread and scattered damages throughout a state."[54] Concentrated damages are far more visible and may be an indication of significant damage to infrastructure supporting neighborhoods and communities, thereby increasing the needs of individuals and families.

However, the dispersion of damage is not necessarily an indication that individual and family needs are non-existent. Damage in rural states, almost by definition, is far less concentrated and could arguably be more difficult for a PDA team to view and assess. Congress has sought to address this challenge through the creation of the Rural and Small State advocate position at FEMA.[55]

## Trauma

Trauma is defined in three ways in FEMA's regulations: the loss of life and injuries, the disruption of normal community functions, and emergency needs that could include an extended loss of power or water.[56]

Despite their prominence and importance to victims, families and communities, the loss of life and injuries have relatively little bearing on a declaration decision, but they would greatly influence media coverage that can influence the decision-making process.

An extreme amount of losses would be traumatic and considered in the evaluation of the governor's request. But the actual help available from FEMA in response to such losses is limited. The Other Needs Assistance (ONA) program, a part of the Individuals and Households Program (IHP), may provide assistance for uninsured funeral expenses and medical help.[57]

As noted, the extreme loss of life or many injuries will likely influence the amount of media coverage for the event. As previously explained in the "Presidential and Gubernatorial Discretion" section of this report, the media coverage can influence not merely the pace of the decision, but the actual decision itself.[58]

The other areas (disruption of functions and loss of utilities) in the trauma rubric can be a point of dispute in declaration decisions. What constitutes a disruption of normal community functions?

Do road closures that result in the closing of schools equate to a disruption? Another consideration is how long the disruption remains and how, or even if, federal help can alleviate the disruption. Similarly, the emergency needs due to the loss of utilities are defined by the length of time the power or water are not in service. Because characteristics of these factors are not defined in regulations, discretionary judgments are significant aspects of the evaluation of IA needs.

## Special Populations

Special populations are considered by FEMA in assessing a request for Individual Assistance.

FEMA attempts to ascertain information about the demographics of an area affected by a disaster event. Those demographics include the age and income of residents, the amount of home ownership in an area, the effect on Native American tribal groups, and other related considerations to be taken into account. The knowledge of the demographics within the affected area gives FEMA added information to consider regarding trauma and community disruption. Special populations are a factor in the consideration of a governor's request, but the total number of households affected that would be eligible for Stafford Act programs remains the prime consideration.

## Voluntary Agency Assistance

Voluntary agency assistance involves an evaluation of what the volunteer and charitable groups and state and local governments have already done to assist disaster victims as well as the potential help they can offer in the recovery period. FEMA also considers whether state or local programs "can meet the needs of disaster victims."[59] This factor is among the most contentious in the disaster declaration process due to the subjectivity of the assessment.

FEMA's evaluation of local and state capabilities, and the capabilities of the local voluntary community, may vary greatly among catastrophes. There is an expectation that the regional office's relationship and history with the state could provide some of this information. But a cursory look at state programs available to "meet the needs of disaster victims" suggests that few resources are comparable to federal help in intent or in scope. A National Emergency

Management Association (NEMA) survey of its members for state-funded disaster assistance showed that relatively few states provide assistance beyond that authorized in the Stafford Act.

NEMA noted that the "other" category could include a range of programs or funds not necessarily of direct aid to victims, including "local government loans, HAZMAT (hazardous materials) funds, disaster unemployment insurance, and a governor's contingency fund."[60] The report does not define these programs any further, but the small numbers of programs directed at IA or unmet needs suggests that FEMA officials evaluate this area carefully given the lack of information regarding available resources for these programs and the extent of their coverage.[61]

**Table 1. Summarizing data from that survey**

| Assistance Provided | Number of States with These Programs |
| --- | --- |
| Public Assistance Program | 22 |
| Individual Assistance Program | 9 |
| Unmet Need Program | 1 |
| Other | 5 |

Source: NEMA 2010 Biennial Report: Organizations & Funding for State Emergency Management and Homeland Security (Lexington, KY: Council of State Governments) p. 4.

In the same vein, attempts to assess the capacity of local voluntary and charitable groups to handle "unmet needs" caused by a disaster can be challenging and problematic. Part of the challenge is discerning the assistance available from the non-profit, voluntary sector, and if that aid meets the needs created by the disaster event. A problematic aspect of the assessment is similar to the "good deeds" concept addressed in the "hazard mitigation" factor in PA. In that instance, the criteria reward the mitigation work that communities and states had undertaken to lessen the impact of a disaster. Similarly, a strong and efficient charitable sector at the local level that is equipped and funded to address the remaining needs could result in a disaster not being declared and federal supplemental funding not being made available.

Arguably, a community does not base all of its preparedness decisions on the potential of FEMA funding in an extraordinary situation. However, it has been argued that some aspects of the FEMA declaration process could be viewed as disincentives for a sound, local capacity to deliver such assistance.

One analyst believed that this problem underlines the need for clear criteria for smaller disaster events:

> Too low a threshold reinforces the perception that the federal government will always come like the cavalry to rescue states and local governments from their improvident failure to prepare for routine disasters. Lapses in preparedness, response, recovery, and mitigation (to cite the disaster management litany) should not be encouraged by a too readily available bailout by the federal government and taxpayers.[62]

In evaluating the capabilities of local organizations, FEMA seeks to determine the help that actually exists, rather than assistance that might be expected to be in place.

## Insurance Coverage

As noted earlier in the PA section, *insurance coverage* is also an important consideration when FEMA considers a request for Individual Assistance. This is in part derived from the general prohibition in the statute of the duplication of benefits, as follows:

> ... each federal agency administering any program providing financial assistance to persons, business concerns, or other entities suffering losses as a result of a major disaster or emergency, shall assure that no such person, business concern or other entity will receive such assistance with respect to any part of such loss as to which he has received financial assistance under any other program or from insurance or any other source.[63]

This provision does not necessarily result in delayed assistance. FEMA is able to provide help to individuals and households that have disaster damages but are waiting on insurance or other assistance for help. Those applicants can receive FEMA help as long as they agree to reimburse FEMA when they receive their other assistance. If a disaster occurred where almost all of the damaged dwellings were fully insured for the damage that was sustained, FEMA could conclude that a disaster declaration by the President was not necessary. Among the types of disasters FEMA frequently responds to, tornado disasters particularly reflect this challenge since tornado coverage is a part of most homeowners insurance policies.

Since the NFIP is administered by FEMA, officials can quickly determine the status of flood insurance in communities and the number of policies in place in the affected area. Additionally, knowledge of the income of the area's residents, as suggested in the *special populations* factor, allows the agency to make some projections regarding the likelihood of insurance coverage, particularly special hazard insurance such as flood or earthquake insurance, which are potentially expensive additions to a homeowners policy. As one analyst has noted:

> The decision not to buy insurance for earthquakes, floods, hurricanes and other natural disasters isn't always conscious: some homeowners don't realize they're not already covered. But many others, faced with high premiums and policies with limited coverage, gamble that they won't need insurance help to rebuild after a disaster.[64]

**Table 2. Average Amount of Federal Assistance Per Disaster Based on Size of State, July 1995 - July 1999**

| Categories | Small States <2Million Population | Medium States 2-10 Million Population | Large States >10 Million Population |
|---|---|---|---|
| Avg. population (1990 Census Data) | 1,000,057 | 4,713,548 | 15,552,791 |
| Number of disaster housing applications approved | 1,507 | 2,747 | 4,679 |
| Number of homes est. major damage/destroyed | 173 | 582 | 801 |
| Dollar amount of housing assistance | $2.8 million | $4.6 million | $9.5 million |
| Number of Individual and Family Grants (IFG) approved (Now known as ONA) | 495 | 1,377 | 2,071 |
| Dollar amount of IFG assistance (now ONA) | $1.1 million | $2.9 million | $4.6 million |
| Disaster housing/IFG (ONA) combined assistance | $3.9 million | $7.5 million | $ 14.1 million |

Source: 44 CFR §206.48(b) (6).

## Average Amount of Individual Assistance by State

The last factor FEMA considers in assessing IA needs is the *average amount of Individual Assistance by state.* FEMA has issued statistics on average losses but notes that the average numbers used are not a threshold (see *Table 2*). The agency does suggest that the "following averages may prove useful to states and voluntary agencies as they develop plans and programs to meet the needs of disaster victims." The inference is that the levels listed generally are what would be expected in damage to dwellings.

The average numbers that follow are based on disasters that occurred between July of 1995 and July of 1999; the data are 10 years old and of questionable use today. The chart divides states into three categories: small states (under 2 million in population), medium states (2 to 10 million in population) and large states (over 10 million in population). The population amounts are based on the 1990 Census.

While FEMA's regulations stress that these are not thresholds, they are considered by agency officials in determining whether IA will be provided. Presumably states may consider that FEMA help could be forthcoming if damage reaches the state indicator levels. However, since the amounts have not been updated in ten years and are based on a 1990 census, it is difficult to determine the degree to which these numbers are considered. Given Congress's mandate in Section 320 of the Stafford Act, this cannot be an arithmetic formula that solely determines whether assistance is provided. But the presentation of loss indicators may guide states when considering whether to request assistance.

In addition to the issue of the data's currency, a larger and more compelling question is whether the numbers in the chart match up to the Preliminary Damage Assessments (PDAs) for those events. Absent an existing review of detailed information in PDA forms, it is not possible to determine the usefulness of the data in *Table 2*.[65] Still, given the level of experience in this field following literally thousands of disaster declarations over the last 30 years, it could be argued that the numbers of eligible households assisted may be reflective of the estimated damage upon which the decision for a disaster declaration was made.

## Congressional Considerations for the Declaration Process

When Congress considers the mandated GAO reports and other commentary on the declaration process, there are some considerations that Members may wish to review when considering the current declaration process.

## The Composition of Preliminary Damage Assessment Teams

One area of consideration is the composition of Preliminary Damage Assessment teams. Team members are involved throughout the process and include local officials guiding the team, state personnel who assist the governor in requesting assistance, and FEMA staff who work on the disaster if one is declared. While FEMA staff have the opportunity at several levels to refine the information the team gathers on damages and to ask additional questions, the process could be approached in other ways. Some FEMA staff have suggested the formation of several permanent teams that would have PDAs as their prime job task without continuing involvement in particular disasters. This could result in more consistent assessments with the bonus of added perspective of team members with exposure to various disasters in many regions of the country. Establishing independent and expert on-site groups to review a situation is a recommendation that the National Transportation Safety Board (NTSB) has been contemplating in their investigations of transportation accidents.[66]

The PDA is an important part of the declaration process. While it can be subject to challenge, it is also the assessment closest to the event. Given its vital role in the process the PDA deserves close attention to determine if these on-site assessments are accurately reflecting the character of an event and the likely eligible damages following a disaster event.

## Updating and Revising Individual Assistance Averages

As previously noted, the averages that appear in FEMA's declaration process regulations for Individual Assistance (IA) to states are derived from experiences from July 1995 to July 1999.[67] These figures could be updated based on FEMA's more recent experience in delivering this type of assistance.

Legislation introduced in the 111th Congress had a provision that directed FEMA to review the factors that are considered for an IA designation.[68]

While the averages are separated by state population size, there does not appear to be a threshold number equivalent to the PA figure of $1 million in eligible damage. In the case of IA help, a dollar figure may not be desirable; however, numbers of families and households affected, or a minimum number of homes with major damage, could potentially be a starting point in establishing a base IA threshold for consideration of requests.

## Other Potential Disaster Indicators

The current per capita indicator based on state population according to the U.S. Census is clear, but some observers believe it lacks precision. For example, the earlier discussion on PA indicators pointed out the commonality between California and Nevada regarding earthquake risk as well as the growth of the states. Their population sizes are very different but the per capita indicator alone does not necessarily measure a state's fiscal capacity.

The GAO report in 2001 noted that per capita personal income measures do not take into account the taxable income a state may enjoy from businesses or corporations that may have considerable taxable profits. GAO did suggest a different indicator:

> We have previously reported that Total Taxable Resources (TTR), a measure developed by the U.S. Department of Treasury, is a better measure of state funding capacity in that it provides a more comprehensive measure of the resources that are potentially subject to state taxation.[69]

FEMA's response to this suggestion questioned whether the use of TTR would be in violation of the legislative prohibition against arithmetic formulas.[70] The TTR does have very specific information and also tracks growth within a state. However, this appears to make it a more accurate measurement. The degree of detail in the measurement would not necessarily make it the "sole" determinant of disaster aid. It could remain an indicator and one among several factors to be considered in the declaration process.

## Conclusion

The disaster declaration process, though subject to inquiry, argument, hearings, studies and recommendations, has changed very little over time. It remains a process that can be observed and evaluated as it occurs in the area affected by the disaster, and grows opaque as it moves up through layers of FEMA and DHS management to the White House. By making PDA information available, FEMA has begun to lift the veil on the decisions that are made. Congress has demonstrated an interest in this process and has sought to understand its limits and its effectiveness. "From the Government Accountability Office in 1981 to Vice President Gore's National Performance Review in 1994," one student of the process noted, "countless policy critiques have called for more objective criteria for presidential disaster declarations."[71] But those calls can be muted when a disaster occurs in a particular place that has particular importance to actors in the process, whether in Congress or the executive branch.

While criteria have been established to create a relatively uniform and knowable process, these criteria may not be determinative of the most critical elements considered. An emphasis on victims of unexpected natural disaster events will likely always have a compelling influence on the disaster declaration process. But since disaster relief does have a political element, at many levels, precedent arguably can have value in improving the declaration process so that it can be applied broadly and fairly. But precedent can also be problematic when discussing events of very different size and impact.

During the last Congress, attention was directed to the nature of the disaster declaration process. The 112th Congress may choose to consider a broad review of the process that might include the consistency of FEMA's approach across the nation in making damage assessments, other potential indicators of state capabilities and capacities, and the currency of the factors it employs to evaluate those assessments of disaster damage and the state requests on which they are based.

## End Notes

[1] 42 U.S.C. § 5122.
[2] Ibid.
[3] For additional discussion, see CRS Report RL34724, *Would an Influenza Pandemic Qualify as a Major Disaster Under the Stafford Act?*, by Edward C. Liu.

[4] For more information on disaster supplementals, see CRS Report RL33226, *Emergency Supplemental Appropriations Legislation for Disaster Assistance: Summary Data*, by Justin Murray and Bruce R. Lindsay. For a map of disaster declarations, see http://www.bakerprojects.com.fema/Maps/CONUS.pdf.

[5] Rutherford H. Platt, *Disasters and Democracy: The Politics of Extreme Natural Events*, (Washington, DC: Island Press, 1999), p. 66.

[6] 42 U.S.C. § 5170.

[7] 42 U.S.C. § 5191.

[8] Michael J. Trebilcock and Ronald J. Daniels, "Rationales and Instruments for Government Intervention in Natural Disasters" in Ronald J. Daniels, Donald F. Kettl, and Howard Kunreuther, eds., *On Risk and Disaster: Lessons from Hurricane Katrina*, (Philadelphia: University of Pennsylvania Press, 2006), p. 104. For additional discussion of the evolution of emergency management see Claire B. Rubin, ed. *Emergency Management: The American Experience 1900-2005* (Fairfax, VA: Public Entity Risk Institute, 2007).

[9] U.S. Department of Homeland Security, Federal Emergency Management Agency, Annual Report of FOIA Activity for FY2000, II. C., at http://www.fema.gov/doc/help/foiareportfy00.doc.

[10] U.S. Department of Homeland Security, Federal Emergency Management Agency, *Preliminary Damage Assessment Reports*, at http://www.fema.gov/rebuild/recover/pda-reports.shtm.

[11] Ibid.; and Platt, p. 266.

[12] U.S. Government Accountability Office, *Disaster Assistance: Improvements Needed in Disaster Declaration Criteria and Eligibility Assurance Procedures*, GAO-01-837, August, 2001, p. 2.

[13] For more information on emergency supplemental spending for disasters see CRS Report RL33226, *Emergency Supplemental Appropriations Legislation for Disaster Assistance: Summary Data*, by Justin Murray and Bruce R. Lindsay.

[14] U.S. Congress, House Committee on the Budget, *Budgetary Treatment of Emergencies*, p. 68, 105th Congress, 2nd. Sess., June 23, 1998 (Washington: GPO, 1998).

[15] For more information, see CRS Report RL33053, *Federal Stafford Act Disaster Assistance: Presidential Declarations, Eligible Activities, and Funding*, by Keith Bea.

[16] Ibid.; and Platt, p. 10.

[17] U.S. Congress, Senate Committee on Appropriations, Subcommittee on VA, HUD and Independent Agencies, *Departments of Veterans Affairs and Housing and Urban Development and Independent Agencies Appropriations for Fiscal Year 2002*, 107th Congress, 1st sess., p. 252, at http://www.access.gpo.gov/congress/senate.

[18] Opening Statement of Rep. Bennie Thompson, in U.S. Congress, House Committee on Homeland Security, *Disaster Declarations, Where is FEMA in a Time of Need?*, 110th Congress, 1st. sess., March 15, 2007.

[19] P.L. 110-28, 121 Stat. 155.

[20] U.S. Congress, House Committee on Appropriations, *Making Emergency Supplemental Appropriations for the Fiscal Year ending September 30, 2007 and other purposes, report to accompany H.R. 1591*, 110th Cong., 1st sess., March 20, 2007, H.Rept. 110-60, p. 211.

[21] Government Accountability Office, *FEMA Disaster Cost Estimates*, Feb. 2008, p. 4.

[22] Matt Korade, "FEMA's Success Hinges on Regional Personnel," *Congressional Quarterly*, March 16, 2010, at http://homeland.cq.com/hs/display/do?docid=3569657&sourcetype=31.

[23] P.L. 109-295, 689g, 120 Stat. 1453.

[24] For more information see CRS Report RL33729, *Federal Emergency Management Policy Changes After Hurricane Katrina: A Summary of Statutory Provisions*, coordinated by Keith Bea.

[25] U.S. Congress, House Subcommittee on Investigations and Oversight, *The Federal Emergency Management Agency's Proposed Disaster Relief Regulations*, 100th Cong., 1st sess. (Washington: GPO, 1987), pp. 3-5.

[26] 42 U.S.C. 5163.

[27] For more information, see CRS Report RL33090, *Robert T. Stafford Disaster Relief and Emergency Assistance Act: Legal Requirements for Federal and State Roles in Declarations of an Emergency or a Major Disaster*, by Elizabeth B. Bazan.

[28] 42 U.S.C. §5191.

[29] President Carter (Georgia), President Reagan (California), President Clinton (Arkansas), and President George W. Bush (Texas).

[30] Ibid.; and Platt, p. 20.

[31] CFR 44 § 206.38.

[32] CFR 44 § 206.33.

[33] Richard Sylves, Quick Response Report #86: *The Politics and Administration of Presidential Disaster Declarations: The California Floods of Winter 1995*, at http://www.colorado.edu/hazards/research/qr/qr86.html.

[34] 44 CFR §206.33(c).

[35] FEMA has 10 regional offices: Region 1 (Boston, MA), Region 2 (New York, NY), Region 3 (Philadelphia, PA), Region 4 (Atlanta, GA), Region 5 (Chicago, IL), Region 6 (Denton, TX), Region 7 (Kansas City, MO), Region 8 (Denver, CO), Region 9 (Oakland, CA), and Region 10 (Bothell, WA).

[36] Sally Kestin and Megan O'Matz, "FEMA ruled on disaster before verifying Dade damage," *South Florida Sun-Sentinel*, at http://www.sun-sentinel.com/news/sfl-fema15may15,0,6848867.story?coll=sfla-news-utilities.

[37] U.S. Department of Homeland Security, Federal Emergency Management Agency, "Joint State-Federal Preliminary Damage Assessment Teams Visit Flood-Damaged Communities," Release Number R10-06-047, November 13, 2006, at http://www.fema.gov/news/newsrelease.fema?id=31509.

[38] 44 CFR 206.33(d)

[39] Ibid.; and Sylves, p. 5.

[40] U.S. Department of Homeland Security, Federal Emergency Management Agency, *Presidential Declarations: What Does This Mean for Me?*, at http://www.fema.gov/media/archives/2007/010807b.shtm.

[41] U.S. Department of Homeland Security, Federal Emergency Management Agency, "Notice of Adjustment of Statewide Per Capita Impact Indicator," 75 *Federal Register*, 62135, October 7, 2010.

[42] U.S. Census Bureau, *State & County Quick Facts*, at http://quickfacts.census.gov/gfd/states/06000.html.

[43] 42 U.S.C. §5170.

[44] 44 CFR § 206.48.

[45] U.S. Geological Survey, Earthquake Hazards Program, "Earthquake Density Maps for the United States", at http://earthquake.usgs.gov/regional/states/us_density.php.

[46] U.S. Department of Homeland Security, Federal Emergency Management Agency, "Notice of Adjustment of Countywide Per Capita Impact Indicator," 75 *Federal Register*, 62135, October 7, 2010.

[47] For information on public insurance coverage, see "Cost of Risk Survey" conducted by the Public Entity Risk Institute (PERI) at http://www.riskinstitute.org.

[48] 44 CFR §206.253. For additional information, see CRS Report RS22945, *Flood Insurance Requirements for Stafford Act Assistance*, by Edward C. Liu.

[49] 44 CFR §206.252.

[50] 42 U.S.C. § 5170c.

[51] For more information, see CRS Report RS22268, *Repairing and Reconstructing Disaster-Damaged Roads and Bridges: The Role of Federal-Aid Highway Assistance*, by Robert S. Kirk.

[52] For additional discussion of this law, see CRS Report RL34209, *Commercial Fishery Disaster Assistance*, by Harold F. Upton.

[53] For more information, see CRS Report RS21212, *Agricultural Disaster Assistance*, by Dennis A. Shields and Ralph M. Chite.

[54] 44 CFR §206.48(b)(1).

[55] P.L. 109-295, §689g, 120 Stat. 1453.

[56] 44 CFR §206.48(b)(2).

[57] 42 U.S.C. § 5174.

[58] Richard Sylves, Quick Response Report #86: *The Politics and Administration of Presidential Disaster Declarations: The California Floods of Winter 1995*, at http://www.colorado.edu/hazards/research/qr/qr86.html

[59] 44 CFR §206.48 (b) (4).

[60] National Emergency Management Association (NEMA) 2006 Biennial Report: Organizations and Funding for State Emergency Management and Homeland Security, (Lexington, KY: Council of State Governments).

[61] For more information on state assistance programs, refer to the following for summary information on state statutory authorities: CRS Report RL32287, *Emergency Management and Homeland Security Statutory Authorities in the States, District of Columbia, and Insular Areas: A Summary*, by Keith Bea, L. Cheryl Runyon, and Kae M. Warnock.

[62] Ibid.; and Platt, p. 65.

[63] 42 U.S.C. §5155.

[64] Liz Pulliam Weston, "Do You Really Need Disaster Insurance?," *MSN Money*, at http://moneycentral.msn.com/content/Insurance/Insureyourhome/P59648.hsp?Printer.

[65] This statement remains accurate. However, since FEMA has begun to list PDA information on its website, it should be possible to begin to assess the efficacy of the factors employed to determine IA damage and declaration criteria.

[66] For more information, see CRS Report RL33474, *Reauthorization of the National Transportation Safety Board (NTSB)*, by Bart Elias.

[67] 44 CFR § 206.48.

[68] H.R. 3377, Title III, Sec.304.

[69] U.S. Government Accountability Office, *DISASTER ASSISTANCE: Improvement Needed in Disaster Declaration Criteria and Eligibility Assurance Procedures,* GAO-01-837, August 2001, p.11.

[70] Ibid., p. 49.

[71] Ibid.; and Platt, p. 285.

# INDEX

## #

9/11, 98
9/11 Commission, 98

## A

abatement, 42, 44
abuse, 95
access, 33, 41, 116, 161, 173
accommodation, 74
accountability, 139
accounting, 38, 101, 102, 104
ACF, 38, 39, 40, 102
adjustment, x, 26, 87, 90, 93, 94, 114, 120
Administration for Children and Families, 12, 38, 40, 102
administrators, 116, 117
Afghanistan, 132
age, 38, 74, 165
agencies, vii, viii, ix, x, xi, 2, 8, 10, 14, 19, 20, 32, 38, 42, 45, 46, 60, 61, 64, 66, 76, 77, 86, 108, 110, 113, 114, 116, 118, 119, 125, 132-134, 136, 140, 145, 169
Agricultural Research Service (ARS), 11, 16
air quality, 75
Alaska, 96, 98, 104, 138
ambient air, 109

American Recovery and Reinvestment Act, 110
American Recovery and Reinvestment Act of 2009, 110
American Red Cross, viii, 2, 47, 75, 131, 143, 157
appointees, 155
Appropriations Act, 24, 27, 51, 60, 66, 67, 69, 71, 82, 89, 102, 103, 125, 129, 131
Appropriations Committee, 40, 102, 125, 151
arithmetic, 91, 92, 93, 154, 160, 169, 171
arrest, 60
ARS, 16
arson, 64
assessment, 78, 90, 92, 109, 115, 150, 155, 157, 161, 162, 165, 166, 170
assets, 41, 61, 63, 81, 92, 146
ATF, 64, 67, 106
Attorney General, 59
authority(s), vii, x, xi, 9, 15, 19, 20, 24, 26, 29, 30, 31, 42, 44, 45, 50, 52, 54, 55, 57, 75, 76, 87, 88, 90, 92, 100, 101, 103, 104, 109, 114, 119, 120, 121, 122, 135, 145, 146, 148, 149, 155, 159, 175
awareness, 22, 117

## B

bacteria, 77, 163

# Index

ban, 66
base, 50, 100, 166, 171
beneficiaries, 30, 73
benefits, 15, 17, 19, 68, 73, 122, 156, 167
BJA, 66
BJS, 66
BOP, 65
breakdown, 59, 74
budget allocation, x, 114, 120
budget deficit, 89, 122, 147
Bureau of Alcohol, Tobacco, Firearms, and Explosives, 64, 67
Bureau of Indian Affairs, 47, 49
Bureau of Justice Assistance (BJA), 66
Bureau of Justice Statistics (BJS), 66
businesses, 25, 81, 91, 156, 158, 171

## C

catalyst, 129, 162
catastrophes, x, 125, 145, 163, 165
catastrophic event, xi, 120, 139, 145, 146, 150, 155
catastrophic failure, 70
categorization, 53
CDC, 42, 44
Census, 99, 160, 168, 169, 171, 174
CERCLA, 76, 109
certificate, 69
CFR, 109, 149, 168, 174, 175
challenges, 158
chemicals, 77
Chicago, 95, 97, 128, 174
childhood, 38
children, 32, 33, 38
China, 98
city(s), 54, 61, 75, 128, 174
citizens, 59, 65, 80
civil action, 59, 76
civil rights, 54, 59, 62
Clean Air Act, 77
Clean Water Act, 76, 79, 109, 110
cleaning, 77
cleanup, 15, 16, 19, 75, 76
Clinton Administration, 138

clothing, 118
CNN, 159
Coast Guard, 12, 46, 48, 49, 76, 109
cognitive development, 38
Cold War, 117
commercial, 24, 163
commodity, 17, 19
community(s), viii, ix, xi, 2, 7, 21, 22, 23, 25, 41, 42, 50, 51, 52, 54, 55, 65, 79, 84, 89, 95, 100, 104, 113, 116, 141, 146, 147, 151, 154, 156, 164, 165, 166, 168
competitive grant program, 68
compilation, 69, 125
compliance, 79, 109
complications, 158
composition, 170
Comprehensive Environmental Response, Compensation, and Liability Act, 109
conference, 31
Conference Report, 100, 105, 106, 107
conflict, 92
congress, 173
Congressional Budget Office, 143
consent, 118
Consolidated Security, Disaster Assistance, and Continuing Appropriations Act of 2009, 132
Constitution, 80, 148, 155
construction, viii, 2, 29, 30, 31, 39, 40, 69, 73, 74, 79, 108, 133, 135, 149
contaminant, 76
contingency, 30, 137, 138, 166
controlled substances, 63
Controlled Substances Act, 63
controversial, 137
cooperation, 61, 63, 157
cooperative agreements, 109
coordination, 8
corruption, 62
cost, ix, 3, 26, 51, 52, 53, 57, 58, 65, 73, 82, 83, 84, 86, 88, 89, 93, 94, 104, 110, 120, 121, 132, 137, 138, 139, 148, 150, 151, 152, 155, 158, 159, 160, 162
cost-benefit analysis, 162
counsel, 59

# Index

counseling, 26, 40, 41, 115, 158, 163
counterterrorism, 125
covering, 74
crimes, 61, 62
criminal justice system, 66
crop(s), 16, 17, 63
CT, 143, 144
currency, 147, 169, 172
Customs and Border Protection, 12, 46, 48, 49
cyclones, 96

## D

damages, 3, 5, 6, 7, 25, 77, 81, 83, 84, 90, 93, 129, 131, 142, 158, 161, 162, 163, 164, 167, 170
danger, 61
database, 14
DEA, 63, 64, 106
deaths, 5, 6, 96
decision makers, xi, 146
decision-making process, 164
deficit, 122, 137, 139, 140
denial, 6, 157
dental care, 30
Department of Agriculture, 15, 16, 46, 48, 49
Department of Commerce, 11, 24, 46, 48, 100, 134
Department of Defense, viii, 2, 28, 30, 31, 46, 48, 49, 60, 61, 62, 64, 65, 67, 68, 71, 99, 100, 105-107, 110, 130, 133, 142
Department of Education, ix, 3, 32, 36, 37, 101, 132
Department of Energy, 46, 48, 49, 89, 100
Department of Health and Human Services, 38, 41, 46, 48, 49, 102
Department of Homeland Security, vii, viii, x, 2, 12, 44, 46, 48, 49, 89, 103, 109, 110, 111, 114, 122, 142, 156, 173, 174
Department of Justice, 47, 48, 49, 59, 60, 61, 63, 64, 65, 66, 67, 105, 106, 107
Department of Labor, 47, 48, 49, 68, 69

Department of the Interior, 12, 47, 48, 49, 130
Department of the Treasury, 47, 64, 91
Department of Transportation, 13, 47, 48, 49, 70, 72, 109, 126, 133, 135
depression, 4
depth, 111
destruction, vii, 1, 3, 10, 132, 133
detainees, 61
DHS, vii, x, 90, 91, 109, 114, 133, 142, 156, 172
directives, 46
disappointment, 156
disaster area, 24, 70, 142
disaster assistance, viii, 2, 6, 8, 9, 10, 15, 24, 26, 27, 36, 50, 70, 83, 84, 86, 89, 92, 95, 103, 116, 117, 118, 120, 125, 130, 131, 132, 133, 136, 138, 142, 143, 144, 147, 148, 155, 162, 166
Disaster Relief Fund, v, vii, viii, ix, 2, 10, 14, 16, 20, 25, 28, 31, 36, 40, 44, 45, 57, 58, 67, 69, 72, 80, 85, 102, 103, 110, 111, 113, 115, 116, 124, 142, 147, 150, 152, 156
disbursement, 139
disinfection, 77
dislocation, 25, 68, 107
dispersion, 164
distress, 25, 26
distribution, 19, 54, 63, 64, 71, 77, 78
district courts, 110
District of Columbia, 33, 35, 44, 69, 141, 143, 175
DNA, 62, 106
DOJ, 59
DOL, 68, 69
DOT, 70
draft, 154
DRF, vii, viii, ix, x, 2, 10, 41, 45, 51, 57, 84, 87, 89, 103, 110, 113, 114, 115, 116, 118, 119, 120, 121, 122, 123, 125, 129, 130, 132, 133, 136, 137, 138, 141, 142, 143, 147, 152, 156
drinking water, 76, 77, 79
drought, 163

drug abuse, 66
Drug Enforcement Administration (DEA), 13, 63, 67, 106
drug trafficking, 63
drugs, 41, 63
drying, 77

## E

earthquakes, 158, 161, 168
Easter, 96
economic development, 24, 54, 131
economics, 92
education, 32, 34-37, 40, 100, 101, 132
El Niño, 128
elected leaders, 154
Emergency Assistance, x, 11, 18, 26, 50, 83, 99, 100, 104, 105, 115, 117, 118, 126, 127, 128, 129, 145, 146, 174
emergency declarations, 44, 92, 94, 111, 146
emergency management, 6, 45, 84, 117, 173
emergency preparedness, 119
Emergency Relief Program, 70, 108
emergency response, 25, 76, 79, 109, 133
employees, 157
employment, 68, 107
energy, 76
enforcement, 60, 63, 122, 123
engineering, viii, 2
enrollment, 32, 34, 35
environment, 59, 76, 95, 109
environmental contamination, 75
environmental impact, 147
environmental protection, 22
Environmental Protection Agency (EPA), 13, 47, 48, 49, 75, 76, 77, 78, 97, 108, 109, 110, 134
environmental standards, 105
Equal Employment Opportunity Commission, 47
equipment, 29, 30, 31, 34, 42, 44, 74, 79, 80, 81, 103
equity, xi, 146, 149
erosion, 19

espionage, 62
ETA, 68, 69, 107
evacuation, 5, 70
evolution, 173
execution, 148, 152, 155
executive branch, 59, 91, 116, 149, 172
executive order(s), 109
exercise, 148, 155, 156, 157
expenditures, 8, 10, 15, 40, 50, 86, 95, 110, 119, 131, 141, 143, 148, 150, 156
explosives, 64
exposure, 170

## F

FAA, 70, 71, 72, 108
fairness, xi, 80, 146
families, xi, 45, 55, 57, 58, 66, 94, 103, 119, 145, 146, 147, 155, 163, 164, 171
family members, 30, 61
Farm Bill, 18
farmers, 15, 16, 18, 130
farmland, 18, 19
federal agency, 8, 28, 76, 139, 167
federal aid, 91, 118, 119, 153, 156, 163
federal assistance, vii, viii, ix, 1, 2, 3, 4, 6, 7, 9, 14, 85, 90, 92, 93, 117, 119, 122, 131, 139, 140, 141, 147, 151, 156, 159, 160, 164
Federal Bureau of Investigation (FBI), 13, 61, 62, 67, 106
Federal Communications Commission, 48
federal courts, 80
federal criminal law, 62, 64
Federal Emergency Management Agency, viii, ix, 2, 10, 12, 44, 45, 53, 55, 75, 89, 95, 96, 103, 110, 113, 115, 141, 144, 147, 173, 174
federal employment, 68
federal facilities, 125, 133
federal funds, 66, 79, 94, 101, 150
federal government, ix, xi, 3, 6, 8, 24, 42, 53, 55, 59, 60, 68, 75, 84, 85, 86, 92, 93, 94, 95, 104, 113, 116, 117, 119, 123, 139, 140, 141, 142, 146, 151, 167

# Index

Federal Highway Administration, 13, 70, 71, 162
federal judiciary, viii, 2
federal law, 59, 60, 75
Federal Register, 103, 174
federal-aid highway system, 70, 71
federalism, 6
FHWA, 70, 72, 162
financial, vii, viii, 1-4, 20, 21, 22, 50, 52, 66, 67, 95, 138, 140, 143, 150, 152, 167
fingerprints, 106
firearms, 64
fires, 22, 128
fish, 24
Fish and Wildlife Service, 13
fisheries, 25
fishing, 24, 25, 163
flexibility, 122
flooding, 5, 6, 34, 88, 126, 128, 140, 143
flood(s), 9, 19, 32, 34, 35, 39, 41, 102, 128, 129, 130, 135, 139, 151, 168
Food, Conservation, and Energy Act of 2008, 18
force, 5, 100
foreign intelligence, 62
forest restoration, 130
formation, 170
formula, 39, 54, 68, 79, 83, 87, 90, 91, 102, 107, 110, 144, 154, 160, 169
fraud, 60, 95
fruits, 17

## G

GAO, 91, 92, 102, 111, 139, 142, 144, 150, 152, 154, 170, 171, 173, 175
GDP, 124, 129
General Services Administration, 13, 47, 48, 49, 135
Georgia, 101, 151, 174
God, 149
good deed, 162, 166
governments, 50, 52, 53, 66, 104, 118, 141, 142, 161

governor, ix, xi, 113, 115, 138, 146, 150, 155, 156, 157, 158, 159, 160, 162, 164, 165, 166, 170
grant programs, 15, 65, 66, 91
grants, 21, 24, 26, 27, 34, 35, 42, 54, 59, 66, 68, 78, 79, 94, 103, 107, 110, 115
grasslands, 20
groundwater, 109
growth, 149, 171
guidance, 157
guidelines, 51
Gulf Coast assistance, vii, 1, 3, 86
Gulf of Mexico, 4, 25, 59, 60, 61, 62, 64, 65, 66, 68, 70, 71, 73, 74, 99, 100, 110
Gulf States, vii, 1, 15, 17, 25, 79

## H

habitat, 25
Hawaii, 101, 138
hazardous materials, 75, 166
hazardous substances, 76, 109
hazardous waste(s), 29, 75, 77
hazards, 19, 45, 118, 174, 175
health, 21, 23, 28, 30, 38-42, 44, 54, 103
Health and Human Services, 134
health care, 30, 41, 42, 44
health care costs, 31, 42, 44
health care system, 42
health condition, 21
health insurance, 103
health services, 39
HHS, 38, 40, 41, 42, 43, 44, 103
high school, 69
high school diploma, 69
higher education, ix, 3, 34, 35, 101
history, 3, 5, 66, 97, 98, 99, 106, 119, 149, 153, 165
home ownership, 165
Homeland Security Act, 64
homeowners, 24, 81, 158, 167, 168
homes, 82, 158, 168, 171
honesty, 123
hostility, 117

182    Index

House, 89, 99, 105, 106, 107, 108, 111, 125, 129, 143, 144, 151, 153, 173, 174
House Committee on Transportation and Infrastructure, 99
housing, viii, 2, 23, 28, 30, 54, 55, 57, 58, 61, 74, 93, 105, 115, 116, 119, 130, 163, 168
Housing and Urban Development (HUD), viii, 2, 47, 48, 49, 54, 55, 56, 57, 58, 99, 105, 130, 131, 132, 133, 134, 173
human, 76
human health, 76
Hurricane Andrew, 97, 151
Hurricane Gustav, viii, 2, 5, 21, 38, 46, 48, 53, 81, 99, 101
Hurricane Ike, viii, 2, 6, 21, 38, 42, 46, 49, 53, 55, 57, 72, 81, 97, 98, 101, 103
Hurricane Katrina, 4, 5, 9, 21, 32, 34-36, 41, 42, 43, 44, 51-53, 58, 67, 69, 73, 74, 82, 83, 88, 98, 99, 102, 103, 108, 119, 126, 127, 130, 131, 133, 138, 146, 173, 174
hurricane relief, viii, 2, 10, 15, 19, 28, 31, 42, 71, 101, 135, 139
Hurricane Rita, 5, 32, 34-36, 53, 83, 84, 99
Hurricane Wilma, 5, 53
hypothesis, 157

## I

ID, 96, 98
identification, 159
Immigration and Customs Enforcement, 12, 47, 48
improvements, xi, 71, 100, 139, 146
income, 38, 54, 55, 58, 69, 91, 105, 140, 154, 165, 168, 171
individuals, ix, xi, 39, 41, 42, 45, 68, 69, 74, 84, 92, 94, 95, 99, 103, 113, 115, 118, 145, 146, 147, 155, 156, 163, 164, 167
Indonesia, 98
industry(s), 21, 24, 25, 64, 68, 160, 163
inflation, 90
information sharing, 62

infrastructure, xi, 24, 25, 29, 30, 70, 71, 76, 79, 80, 94, 100, 109, 130, 145, 146, 154, 159, 161, 162, 163, 164
injury(s), viii, 2, 5, 82, 164
inmates, 65
insects, 21
inspectors, 115
institutions, 34, 35, 101
instructional materials, 34
intelligence, 62, 63
international law, 60
international trade, 100
investments, 139, 161
Iowa, 101
Iraq, 66, 67, 71, 82, 132
issues, ix, 3, 66, 113, 115, 116

## J

Job Corps, 13, 68, 69, 108
job creation, 26
job training, 26, 68, 69
job training programs, 68
judiciary, 80
jurisdiction, 63, 76
justification, 106
juvenile delinquency, 66
juvenile justice, 66

## L

landscape, 149
large-scale disasters, 122
law enforcement, 59, 60, 62, 67, 125
laws, 9, 19, 32, 35, 51, 63, 76, 109, 117, 148, 155
laws and regulations, 63
layoffs, 68
lead, 6, 61, 70, 76, 108, 109, 152, 157
leadership, 62, 66
learning, 34
legislation, vii, x, xi, 11, 39, 68, 82, 84, 88, 95, 100, 111, 114, 116, 119, 122, 125,

# Index

129, 132, 133, 136, 139, 142, 145, 146, 150, 154
light, 95, 140, 141
liquidate, 102
livestock, 16, 17, 18
loan guarantees, 24, 104
loan principal, 53
loans, ix, 2, 23, 50, 51, 52, 53, 79, 81, 82, 92, 94, 104, 118, 132, 166
local authorities, 62
local community, 66
local government, xi, 6, 45, 50, 52, 53, 64, 75, 76, 77, 90, 91, 94, 103, 104, 115, 116, 145, 147, 148, 151, 158, 159, 161, 165, 166, 167
Love Canal, 97

## M

machinery, 81
magnitude, 6, 116, 147, 148, 156
major disaster declaration, ix, xi, 3, 52, 53, 85, 90, 92, 104, 110, 113, 116, 129, 133, 145, 146, 147, 148, 155, 163
majority, 7, 34, 35, 38, 68, 70, 74, 79, 131, 138, 150
man, 45, 147
management, x, 25, 46, 63, 75, 77, 114, 117, 118, 152, 167, 172
man-made disasters, 45
manufactured housing, 119
manufacturing, 100
mapping, 25
marine fish, 24
marketing, 21
Maryland, 80, 101
mass, 68, 118, 147
mass care, 118
mass media, 147
materials, 29, 149
MB, 122
measurement(s), 92, 160, 171
media, 157, 159, 164, 174
Medicaid, 12, 41, 42, 102
medical, 30, 31, 41, 73, 74, 164

medical care, 73, 74
Medicare, 12, 42
mental health, 39, 40, 41, 42
methodology, x, 114, 119, 120
metropolitan areas, 54
Mexico, 65, 67, 68, 71
Midwest flooding, 88
military, viii, 2, 28, 29, 30, 100, 135
mission, viii, 2, 16, 41, 45-49, 57, 63, 65, 70, 71, 76, 77, 80, 105-108, 152
Mississippi River, 12
Missouri, 101
modernization, 29
modifications, 79
momentum, 159
moratorium, 23
motivation, 157

## N

narratives, 8
National Aeronautics and Space Administration, 14, 47
National Crime Information Center, 62, 106
national debt, 89
national emergency, 107
National Forest System, 20
National Park Service, 13, 47, 48
National Response Framework, 21, 41, 102, 109
national security, 125
Native Americans, 130
natural disaster(s), 15, 17, 18, 19, 20, 24, 32, 34, 35, 36, 39, 45, 68, 70, 77, 101, 117, 122, 129, 149, 150, 162, 168, 172
natural resources, 19, 20, 59, 76
Natural Resources Conservation Service (NRCS), 11, 16, 19
NCA, 73
NCP, 75, 76
needy, 74
New Deal, 149
NIJ, 66
No Child Left Behind, 101
NOAA, 24, 25

nonprofit organizations, xi, 45, 69, 82, 145, 161
Northridge Earthquake, 97, 98, 128
NRCS, 19, 20
NRF, 21, 70, 76
nutrition, 17, 19

# O

Obama, 122
Obama Administration, 122
objective criteria, 151, 156, 172
offenders, 65, 66
Office of Justice Programs (OJP), 13, 65, 67, 107
Office of Management and Budget, x, 16, 86, 114, 119, 120, 123, 142, 144
Office of the Inspector General, 30
officials, 22, 63, 76, 77, 99, 116, 118, 136, 153, 158, 159, 166, 168, 169, 170
oil, 76, 77, 88, 109, 126, 130
Oil Pollution Act (OPA), 76
oil spill, 76, 88, 109, 126, 130
OJP, 65, 66, 67
Oklahoma, 128
OMB, x, 86, 114, 119, 120, 122, 123, 138, 139, 141, 143
omission, 122
OPA, 109
operations, viii, 2, 29, 30, 33, 35, 46, 60, 62, 66, 74, 129, 152
opportunities, 65
organize, 22
outreach, 99
oversight, vii, 1, 4, 58, 86, 95, 119, 139
overtime, 95
oyster, 25

# P

P.L. 105-220, 68
*Pandemic Influenza Act*, 60, 61, 62, 64, 65, 67, 68, 71, 99, 100, 110
parents, 33
participants, 69, 157
payroll, 74
per capita income, 140
permit, 50
personal communication, 25
personnel costs, 28
persons with disabilities, 57
petroleum, 109
PHAs, 55, 57, 58
Philadelphia, 74, 173, 174
plants, 21, 77, 79
police, 23
policy, 86, 89, 117, 136, 137, 157, 168, 172
policy options, 136
policymakers, 117
pools, 103
population, 8, 24, 39, 87, 91, 117, 154, 160, 168, 169, 171
population growth, 87
population size, 171
poverty, 39
precedent(s), 118, 149, 172
preparedness, 45, 74, 94, 136, 166, 167
President Clinton, 174
presidential declaration, 90, 115, 147, 154, 158, 162, 163
price index, 154
primacy, 154
principles, 6
prisoners, 61
prisons, 65
private sector, 26
procurement, 29
producers, 15, 17
product market, 25
professionals, 66, 159
profit, 38, 81, 118, 139, 159, 166
program administration, 25
program staff, 161
project, 26, 27, 69, 71, 73, 75, 78, 107, 122
property taxes, 50
protection, 21, 28, 45, 50, 60, 109, 146
prototypes, 29
public assistance, 83, 84, 119, 154
public education, 33

# Index

public employment, 68
public health, xi, 42, 50, 76, 78, 92, 145, 146
public housing, 55, 57, 58
public law, viii, 2, 9, 32
public safety, 159
public schools, 33
public sector, 156
public service, 26, 50
Puerto Rico, 54, 69

## Q

quality of life, 24

## R

Radiation, 78, 79
reactions, 153
real estate, 81, 82
real property, 30
recognition, 84
recommendations, xi, 90, 92, 146, 149, 150, 152, 154, 156, 172
reconciliation, 11
reconstruction, 70, 71, 78
recovery plan, 55, 57
recovery process, 50
recruiting, 32, 33
reform(s), 51, 104, 119, 138, 150, 153
regulations, 26, 51, 54, 75, 82, 92, 104, 149, 153-158, 160, 164, 165, 169, 170
regulatory changes, 154
rehabilitation, 15, 18, 69, 99, 100
rehabilitation program, 15, 99, 100
reimburse, 80, 92, 94, 167
relief, viii, ix, x, 2, 10, 15, 19, 23, 24, 28, 31, 36, 42, 55, 71, 74, 87, 95, 101, 113, 114, 117, 119, 125, 130, 132, 135, 138, 139, 140, 141, 149
rent, 55, 58
rent subsidies, 55
repair, 29, 30, 31, 39, 58, 65, 69, 70, 71, 79, 81, 125, 129, 163

requirements, 77, 79, 101, 105, 147, 148, 154
resolution, 11, 59
resources, vii, 1, 4, 20, 45, 84, 89, 91, 92, 95, 108, 116, 139, 148, 149, 152, 155, 158, 165, 166, 171
responsiveness, 157
restoration, 15, 19, 25, 70, 79, 130, 133
restrictions, 90
restructuring, 136, 137
retirement, 28
revenue, 50, 137
rights, 80
risk(s), 45, 93, 103, 162, 171
Robert T. Stafford Disaster Relief and Emergency Assistance Act, x, 26, 50, 83, 105, 117, 118, 145, 146, 174
Robert T. Stafford Emergency Relief and Disaster Assistance Act, ix, 113
rules, 19
runoff, 19
rural areas, 22, 24
Rural Utilities Service (RUS), 16, 24
RUS, 24

## S

safety, xi, 41, 50, 61, 145, 146
Salvation Army, 75
SAMHSA, 42, 43
savings, 137, 162
scaling, 139
schedule delays, 29
school, 32, 33, 34, 36, 38, 50, 101, 104, 132, 165
scope, 6, 149, 158, 165
seafood, 25
secondary education, 36, 37
secondary schools, 32, 33
Secret Service, 12, 47
Secretary of Commerce, 26
security, 60, 80
seizure, 63
self-improvement, 65
Senate, 50, 98, 104, 142, 143, 151, 173

# 186 Index

September 11, 98, 158
services, 38, 39, 40, 41, 42, 45, 50, 60, 62, 68, 69, 70, 73, 95, 108, 110
sewage, 79
Small Business Administration, viii, 2, 8, 14, 55, 81, 82, 110, 132, 135, 157
small businesses, 82
SNAP, 19
social benefits, 53
Social Security, 47, 102
Social Security Administration, 47
social services, 39
social welfare, 149
society, 65
socioeconomic status, 8
Solicitor General, 59
solution, 154
Soviet Union, 117
spending, x, xi, 40, 86, 89, 110, 114, 120, 122, 123, 136, 141, 146, 147, 150, 151, 152, 173
Spring, 96
stabilization, 44
standardization, 86
statistics, 169
statutes, viii, ix, 2, 10, 11, 75, 85, 88, 113, 117, 124, 138, 142
statutory authority, 118
stock, 91
storage, 77, 109
storms, vii, ix, 2-5, 84, 88, 95, 96, 102, 126
stress, 169
stretching, 4
structure, 115, 136, 142, 161
subjectivity, 165
subsidy, 104
substance abuse, 42
Substance Abuse and Mental Health Services Administration (SAMHSA), 41
substitution, 63
sugarcane, 17
Sun, 174
Superfund, 76, 109
supervision, 59

Supplemental Nutrition Assistance Program, 19
suppression, 131
Supreme Court, 59
surplus, 17
surveillance, 41

## T

tanks, 77, 78, 109
TAP, 18
target, 86, 91, 121
Task Force, 136, 137, 142, 143
tax base, 160
taxation, 91, 171
taxpayers, 167
TCR, 97, 98, 99
team members, 159, 170
teams, 41, 158, 170
technical assistance, 18, 20, 21, 24, 26, 107, 130
technical support, 66
technology, 29, 62, 66
telecommunications, 24
temporary housing, 57
tenants, 24
Tennessee Valley Authority, 47, 48, 49
territorial, 62
territory, 147
terrorism, 45
terrorist attack, 62, 158
Terrorist attack, 88, 127
testing, 25, 77
threats, 22, 45, 62, 76
timber production, 22
Title I, 34, 35, 54, 60, 61, 62, 64, 65, 66, 101, 135, 175
Title II, 60, 135, 175
Title IV, 34, 35, 67
Title V, 35, 89, 100, 142
tobacco, 64
tornadoes, 6, 151, 162
toxic substances, 75
tracks, 117, 171
trafficking, 64

training, 29, 30, 40, 63, 66, 68, 69
training programs, 66
transcripts, 143
transparency, 10, 139, 141, 153
transport, 108
transportation, 30, 70, 108, 170
transportation infrastructure, 70
Transportation Security Administration, 46, 48, 49
trauma, 164, 165
Treasury, 48, 49, 171
treatment, 76, 79, 149
triggers, 115
Trust Fund, 74, 78, 135

# U

U.S. Army Corps of Engineers, 28
U.S. Department of Agriculture (USDA), 15, 16, 18, 163
U.S. Department of Labor, 107, 108
U.S. Geological Survey, 13, 47, 49, 97, 174
U.S. history, 6
U.S. Treasury, 38, 40
unemployment insurance, 68, 166
uniform, 149, 172
uninsured, xi, 119, 145, 164
United, 3, 6, 12, 44, 60, 61, 62, 63, 74, 75, 76, 80, 86, 96, 97, 98, 100, 105, 110, 116, 117, 121, 125, 148, 155, 174
United Nations, 63
United States, 3, 6, 12, 44, 60, 61, 62, 63, 74, 76, 80, 86, 96, 97, 98, 100, 105, 110, 116, 117, 121, 125, 148, 155, 174
United Way, 75
universities, 75
urban, 22, 54
USDA, 15, 16, 17, 19, 20, 99, 100
USGS, 97

# V

variables, 149, 157

variations, 149
vegetables, 17
vehicles, 29
vein, 166
Vice President, 172
victimization, 66
victims, 66, 71, 87, 89, 99, 116, 149, 164, 165, 166, 169, 172
violence, 66
violent crime, 62, 67
Volunteers, 75
voting, 122
vouchers, 55, 57, 58, 99, 105

# W

waiver, 42, 83, 84, 101, 103, 109
war, 82, 117, 130
War on Terror, 60, 69, 103
Washington, 44, 61, 74, 98-100, 105-108, 111, 142, 143, 144, 157, 173, 174
waste, 24, 77, 95
waste disposal, 24
wastewater, 76, 77, 79
water, 19, 24, 28, 30, 41, 75, 76, 79, 109, 164, 165
water quality, 75
water resources, 28
watershed, 19
weapons, 29, 117
websites, 8
welfare, 54
White House, 98, 115, 119, 149, 156, 172
WIA, 68, 69, 107
wildfire, 131
wildland, 135
witnesses, 60
workers, 68
workforce, 41, 68
Workforce Investment Act, 68
World War I, 116, 117
worldwide, 61